The purpose of this publication is to provide readers with the most complete history of combined aircraft designs created by Vladimir Vakhmistrov. The designer himself referred to his projects as 'Zveno Aircraft' (where the Russian word 'Zveno' stands for 'Chain Link' or 'Flight' combat unit) – adding the names of aircraft which were included in each configuration. Vakhmistrov was among the world's pioneers in devising and implementing the attachment of small fighters under heavy bombers – thus the larger aircraft were able to carry the smaller ones to the target area, whereas the fighters, in their turn (after the separation from their 'motherships'), acted as guards for the slow-speed giants. During 1931 through to 1941, several configurations of this concept were created – involving mass-produced TB-1 and TB-3 bombers, and I-4, I-5, I-Zet and I-16 fighters which were modified correspondingly in order to fit the task.

Based upon Vakhmistrov's projects, the first in-the-world air connections of fighters under their bomber carriers were carried out, with the most successful variant among all the created 'Zveno Aircraft' consisting of a four-engined Tupolev TB-3 bomber and two Polikarpov I-16 fighters attached under the bomber's wing. The fighters carried two 250kg bombs each and acted as high-speed dive bombers after disconnection from the carrier plane. Several such aircraft – referred to as 'Zveno-SPB' – were used operationally during the war by the Black Sea Fleet Air Force. Initially, three TB-3s with AM-34RN engines were allocated for re-equipment into Zveno-SPB, which were taken into service by the 2nd Special Squadron of the 32nd Fighter Regiment, 62nd Aviation Brigade of the Black Sea Fleet Air Force stationed in Eupatoria. The order to begin combat application of the 32nd Regiment's combined aircraft was given on 22 July 1941, while the first air raid on port structures in Romania – the German ally – was made on 26 July. More than 30 combat missions were carried out until October 1941 and became one of the most successful operations of the Soviet aviation during the Great Patriotic War. Meanwhile, Vakhmistrov had a large number of other interesting projects – and many of them remained on the designer's drawing board.

More than 20 years were spent by the author on the historical research of the subject in the Russian archives, and the book itself is extensively illustrated by pictures from the author's own archive, as well as aircraft test reports and specially prepared scale drawings and color side views.

Mikhail Maslov was born into the family of a serviceman in Mary (USSR, now Turkmenistan) in 1954 and was brought up in a town in the Urals. He graduated from the Moscow Aviation Institute (University) in 1982 with a degree in Mechanical Engineering. Mikhail worked as a structural design engineer at the Tupolev Design Bureau from 1982-1986 and then transferred to the TsAGI engineering information department, where he took various posts – including that of a deputy head of the department. From the end of the 1990s, he was actively involved in the restoration of historical aircraft – including airworthy Po-2, I-15bis, I-153, I-16, DIT and MiG-3 airplanes.

Already during his work at Tupolev, Mikhail became interested in the history of Soviet aviation from the 1920-1940 period and began the process of researching and analyzing archive material. In the last 30 years of his enthusiastic approach to his work, he has written numerous articles and dozens of books on the history of Soviet aircraft of that period.

VAKHMISTROV'S CIRCUS

Zveno Combined Aircraft – The Projects, Development, Testing and Combat

Mikhail Maslov

Helion & Company

Helion & Company Limited
26 Willow Road
Solihull
West Midlands
B91 1UE
England
Tel. 0121 705 3393
Fax 0121 711 4075
Email: info@helion.co.uk
Website: www.helion.co.uk
Twitter: @helionbooks
Visit our blog at http://blog.helion.co.uk/

Published by Helion & Company 2017
Designed and typeset by Mach 3 Solutions Ltd (www.mach3solutions.co.uk)
Cover designed by Paul Hewitt, Battlefield Design (www.battlefield-design.co.uk)
Printed by Hobbs The Printers Ltd, Totton, Hampshire

Outside front cover: Photograph – 23 March 1935: The I-Zet fighter has just become attached to the TB-3 carrier aircraft for the first time. This photo was taken from the accompanying R-5 reconnaissance aircraft. Upper colour profile – Zveno-1, ground testing, 1931; lower colour profile – Zveno-SPB – TB-3 4M-34RN + two I-16s Type 5, NII VVS, November 1938.
Inside front cover: Top – Zveno-1 – ground testing, 1931; centre – Zveno-2 – full configuration: TB-3 4M-17 + three I-5s, 1933; bottom – Zveno-2 – TB-3 4M-17 + two I-5s, 1935.
Inside rear cover: Top – Zveno-5 – TB-3 4M-17 + I-Zet; centre – Aviamatka PVO (Airborne Mothership (A-M) for Air Defense) – TB-3 4M-17 + two I-5s + two I-16s + I-Zet, November 1935; bottom – Zveno-6 – TB-3 4M-17 + two I-16s Type 4, NII VVS, summer 1936.
Outside rear cover: Top – Zveno-7-1 – TB-3 4M-17 + I-16 Type 5, NII VVS, July 1938; bottom – Zveno-SPB – TB-3 4M-34RN + two I-16s Type 5, NII VVS, November 1938.

ISBN 978-1-911096-75-7

British Library Cataloguing-in-Publication Data.
A catalogue record for this book is available from the British Library.

Contents

Introduction

A 'circus' was the unofficial nickname under which the history of Soviet aviation remembers the development of combined aircraft by Vladimir Sergeyevich Vakhmistrov. The designer himself referred to his projects as the 'Zveno Aircraft', where the Russian word 'Zveno' stands for 'Chain Link' or Flight' combat unit – adding the names of aircraft which were included in each configuration.

Vakhmistrov had a lot of other most interesting projects – among them several that were actually implemented. All of these original projects, both implemented and non-implemented, are described in the text below. What comes first, however, is some brief information about the creator of these unique flying vehicles.

VLADIMIR SERGEYEVICH VAKHMISTROV (1897-1972)

After graduation in 1916 from the Mikhailov Artillery School in Petrograd, Vakhmistrov joined the ranks of a mortar division as a warrant officer – and shortly afterwards, went with his division to the front. During a military operation, Vakhmistrov was contused and was sent to Moscow for treatment.

In 1918, Vakhmistrov enlisted in the Red Army as a volunteer; first, he served in an ordnance unit, but then insisted that he be transferred to the Pskov Air Group, where his brother – Boris Sergeyevich Vakhmistrov – was in command of a detachment. As an observer pilot, Vakhmistrov took part in the Civil War and performed missions in the Volga Region and in Turkestan.

In 1922, Vakhmistrov entered the N.E. Zhukovsky Air Force Academy in Moscow as an attendee. At the time, the institution was better known as the Air Fleet Academy. Like many active aviators of the day, military student Vakhmistrov went in for glider flying. From 1923, he was a permanent member of the Technical Committee for the glider-flying section and personally took part in the creation of several gliders, which were designed and constructed by the Air Fleet Academy.

Below are some key dates relating to those projects:

- In 1923, Vakhmistrov took part in the construction of the AVF-1 'Arap' ('Moor') glider, which was designed by M.K. Tikhonravov. This was the first flying vehicle of its type constructed by the attendees in the working shops of the academy.
- In 1924, Vakhmistrov designed a glider of his own – the AVF-8 'Condor' – which was built in the car repair shops within the depot complex of the Council of People's Commissars (*Sovnarkom*).
- In 1925, Vakhmistrov and Tikhonravov designed the 'Zmey Gorynych' ('Firedrake') high-wing monoplane. It should be noted that the aircraft described above did not demonstrate any exceptionally successful flying at gliders' competitions in 1923 and 1924; however, they provided a positive designing experience for everybody who was involved in their creation.
- In 1927, Vakhmistrov took part in the construction of the 'Zhar Ptitsa' ('Firebird') glider, which was designed by Tikhonravov, Dubrovin and Vakhmistrov.
- In 1928, he worked together with B.I. Cheranovsky on the project of the Cheranovsky BICH-7 'parabola aircraft'.
- In 1929, the 'Scythian' glider appeared.
- By 1930, Tikhonravov, Vakhmistrov and A.A. Dubrovin had already become rather famous glider designers.

In 1927, after graduation from the academy, Vakhmistrov was employed by the Air Force Special Procurement Department as a specialist for technical acceptance of incoming equipment and outfit. Later on, he joined the team of the *NII VVS* (*Nauchno-Issladovatel'sky Institut Voenno-Vozdushnykh Sil*, or the Scientific Test Institute of the Air Force), where he took part in tests of new aircraft. The above list of gliders attests to the fact that during this time, Military Engineer Vakhmistrov demonstrated that he was a designer of unique flying vehicles.

In 1930, he designed and built a small 'target glider' for the in-flight firing training of air gunners; in the summer of that year, the glider was installed on the center section of the upper wing of the Polikarpov R-1 reconnaissance aircraft and was successfully tested.

The designer himself described those years of his life as follows:

> In 1930, when I was working for NII VVS, there was once a flying target contest. I made a small glider model, about half a meter in wing span, which was capable of gliding down and pirouetting in the air after disengagement, which added to its value as a target for an air gunner. I designed the model myself, and tested it together with pilot S. Rybalchuk. The model was attached to the Polikarpov R-1 reconnaissance aircraft, and was detachable, whenever necessary, via a linkage from the observer pilot's seat. When we were performing the flight test, I was the one to disengage it …
>
> At the time, I was head of the Heavy Aircraft Test Department at NII VVS. The aircraft under test at the time were the Tupolev TB-1s. Simultaneously, the Tupolev I-4 fighters were also being tested. After successful tests of the flying target it occurred to me that we might attempt to mount the I-4 on the TB-1. Appropriate aerodynamic calculations were performed, and the design was prepared. It was presented for approval to Air Force Commander Y. Alksnis and Chief of Staff S. Mezheninov. They liked the project, and I obtained their approval for implementing it. At the very same time, the concept was first named 'the Zveno Aircraft'.*

In brief, the content of the suggestions made by Military Engineer Vladimir Vakhmistrov were as follows: two Tupolev I-4 fighters were attached to the wing of the Tupolev TB-1 bomber, took off together with it and flew to the designated target. The engines' thrust allowed the TB-1 to lift an increased load and provided for enhanced take-off performance. During the flight, the fighters were refueling from the bomber's fuel tanks – thus their range considerably increased. After the detachment in flight, the fighters acted as a flying covering party for the bomber and escorted it during the flight back to the airfield. The system was named 'the Zveno Aircraft'; in documents of the time, it was referred to as 'SZV', 'SZ-1', 'Z-1' or 'Z-1 Aircraft'. Below, you will find a more detailed description of the events related to the history of the SZ-1 system.

---

* Archive of M.E. Zhukovsky Memorial Museum: a collection of documents relating to V.S. Vakhmistrov; a record of V.S. Vakhmistrov's interview with M.S. Arlazorov on 7 February 1969.

Vladimir Sergeyevich Vakhmistrov, with three bars on each collar patch and the Order of the Red Star, which was awarded to him for the creation – and the successful testing – of the Zveno-1 flying system; this photo was taken during the 1930s.

The 'Scythian' glider and its creators: designers A.A. Dubrovin, V.S. Vakhmistrov and M.K. Tikhonravov (from left to right), and Test Pilot A.B. Yumashev (the second from the right); Koktebel, 1928.

Arriving at the glider contest in Koktebel, 1928. Sitting from left to right are: Carl Venzlav, V.S. Vakhmistrov, K.K. Artseulov (in the shade), S.V. Ilyushin, D.A. Koshits (in flying goggles and helmet), A.B. Yumashev, A.A. Dubrovin and A.A. Beryozkin.

Military Engineer V.S. Vakhmistrov, aged 29; this has been cropped from the previous photo.

1

'Zveno Aircraft': Tupolev TB-1 + Two Tupolev I-4s (Zveno-1)

As soon as the higher authorities approved the concept of the combined aircraft, the history of its creation – and improvement – began. On 30 June 1930, all design materials and initial calculations were sent by Vakhmistrov to the *TsAGI* (*Tsentralniy Aerogidrodinamicheskiy Institut*, or The Central Aerohydrodynamic Institute) for review and further study. There, specialists of the Design Department of the Experimental Aircraft Engineering Sector – headed by A.N. Tupolev – performed the required analyses of the possibility to attach two fighters to the wing of the heavy bomber, and started to develop the working drawings, as well as manufacturing additional structural assemblies.

On 31 January 1931, the *TsAGI* requested that the *NII VVS* provide them with a Tupolev I-4 fighter in order to carry out all the required improvements and modifications. The *NII VVS* furnished the *TsAGI* with an I-4 (s/n 1502), which – for reasons of apparent wear – had been involved only in ground tests and system trials. For practical flights, other aircraft were used – including two rather new Tupolev I-4s (s/n 1651 and s/n 1655) and Tupolev TB-1 s/n 651. Structural modifications were primarily made to the TB-1. Three 'pyramids' for the mounting of fighters were arranged on each side of the bomber's wing, and two forward pyramids for the attachment of the I-4's landing gear were installed along the axis of the front spar of the TB-1. Each pyramid was fitted with a shaped support (latch), which was to be engaged with the mating component on the fighter's landing gear. The third pyramid – with a release latch – supported the I-4's fuselage near its third load-carrying frame. The fighters were mounted near the bomber's rotating propeller disk, with the longitudinal axis of each I-4 distanced 307mm from the axis of the TB-1's wing joint in the direction of the bomber's longitudinal axis; the small lower wings of the I-4s were removed. To mount the fighters onto the wing, they were pushed up a special ramp with wooden troughs.

The first test of the system named 'Zveno Aircraft 1' (SZ-1) took place on 3 December 1931 at the 17th Air Brigade's airfield in Monino (the Moscow area). First, the engines of both Tupolev I-4 fighters were started and warmed up on ground; then, the ground crew rolled them up a sloped ramp to the top of the TB-1's wing. After connection of all systems – and a thorough inspection and check – runs over the airfield with various engine settings began. At 2:30 p.m., the first flight of the SZ-1 took place, with the total take-off weight of 8,180kg (5,640kg (TB-1) + 2 x 1,270kg (two I-4s)). The crew of the TB-1 consisted of pilot Adam Zalevsky (as pilot-in-command in the left-hand seat), Andrei Sharapov (as co-pilot in the right-hand seat), designer Vakhmistrov (in the aft gunners' first hatch) and engineer Morozov (in the second hatch). The left-hand Tupolev I-4 was piloted by Valery Chkalov and the right-hand Tupolev I-4 was piloted by Alexander Anisimov. The SZ-1 uneventfully took off from the airfield at a speed of 140km/h – and Adam Zalevsky did not notice any vibrations or excessive pressures at the controls during take-off and climb. At an altitude of 1,000m and a speed of 160km/h, the first separation of the fighters from the mothership occurred. The Tupolev I-4 which was piloted by Anisimov separated with its tail upward and without sufficient lift; for this reason, it just flew off the pyramids and hit its wheel against the third spar of the TB-1's wing – thus breaking the corrugated skin. In the course of such an abnormal separation, the fighter's propeller collided with the attachment pyramid, which resulted in damage to the propeller blade and to the pyramid itself. In spite of this slight incident, the experiment successfully ended with a safe landing of all the three aircraft in the Monino airfield.

That was what the events looked like in the official report on the tests of the SZ-1, which was personally signed by Vakhmistrov;[*] however, in his further memoirs, Vakhmistrov described the first test in a somewhat different way:

> It was intended that I should fly in the copilot's seat near the TB-1's pilot-in-command Zalevsky. According to the first version of the project, I was the one to open the latches separating the fighters. We followed this procedure during the training; however, in the very last moment, our superiors ordered that the right hand seat should also be occupied by a pilot. So pilot Sharapov took the right-hand seat, whereas engineer Morozov and I had to occupy the aft cabins.
>
> Such 'shuffling' in the very last moment resulted in a near-disastrous situation. When we were taking off we did not know how the system would behave if the fighters separated non-simultaneously. Of course, we wanted to ensure maximum simultaneity. When Zalevsky ordered disengagement, Sharapov lost control over the situation. Without waiting for Chkalov to separate the tail of his aircraft (the fighters released this latch themselves), Sharapov disengaged the wheels. Thus Chkalov's fighter was hanging in the air, attached at the tail point. Yet, he quickly understood the situation, jerked the aircraft to the right and to the left, separated, and flew away. Anisimov displayed a singular presence of mind as well. He quickly disengaged the tail and took off as was required. Of course, after such an incident the latches were replaced with new ones, which ensured automatic separation.[†]

Test Pilot P.M. Stefanovsky had his own version of the event:

> Immediately before the take-off (which was scheduled to take place from an airfield in the Moscow Area) there came un unexpected order stating that engineer Vakhmistrov should be placed in the forward gunner's cabin, and that the copilot's duties should be performed by Squadron Commander Andrei Rodionovich Sharapov.
>
> The combined aircraft took off uneventfully, and started climbing. At the target altitude and at the strictly designated moment, Zalevsky signaled to Sharapov to disengage the fighters. The procedure was first to release the aft latches to disengage the tails of the I-4s, and then to release the forward latches to let them get a lift of their own and to separate from the bomber. The levers for releasing the forward latches were located in the lower area in the passage between the pilot-in-command's seat and the copilot's seat. So Andrei in a hurry released the forward lock earlier than the aft one.
>
> Chkalov felt at once that though the landing gear was released the tail was still attached. The lift kept growing; now there might be a loop around the attached tail, and... Valery immediately assessed the situation and responded promptly. He pushed the stick, released the aft lock, and, with the fighter's wheels exercising downward pressure on the bomber's upper skin, flew away.
>
> Anisimov's fighter was still on the bomber's wing. In spite of the earlier anticipations, the TB-1 maintained sufficient controllability. Thus, the last test task turned out to be among the first ones to be fulfilled. In the course of disengaging Anisimov's fighter, Sharapov acted already quite cold-bloodedly. The separation was performed successfully.
>
> After the flight, Vladimir Sergeyevich Vakhmistrov for some time tended to treat Andrei

---

* P.M. Stefanovsky, *The Unknown 300* (Moscow, 1968).

† Archive of M.E. Zhukovsky Memorial Museum: a collection of documents relating to V.S. Vakhmistrov; a record of V.S. Vakhmistrov's interview with M.S. Arlazorov on 7 February 1969.

> Sharapov with a certain dislike. He understood, though, that the design should be modified. First of all, he moved the aircraft disengagement mechanism to the fighters' cabins. Then the pilots of the carried aircraft, when receiving the order to separate, could themselves release the tail latches and keep the aircraft attached at two points using the elevator. All they had to do to separate completely was to pull the control stick. The forward latches became released at once, and the fighters separated from the TB-1 and climbed to a dozen of meters, a little bit behind the bomber.*

After the first flight during winter 1931/1932, the entire SZ-1 system was substantially modified. With the help of *TsAGI* engineers Petlyakov, Belyayev, Priven, Nikitenko and Fomin, additional wing strength calculations were performed in order to enhance the efficiency and reliability of the combined aircraft. Furthermore, to increase the SZ-1 take-off weight, an additional ANT-9 shock strut was arranged on the outer side of the TB-1's main landing gear legs. The mothership's wing spar at the attachment point of such a landing gear leg was reinforced using an additional internal brace. Additionally, the TB-1 was fitted with extra-sized 1,350 x 300mm wheels instead of the previous 1,250 x 250 mm ones. The third pyramid attaching the fighters' fuselages was now attached to the TB-1 wing using a hinged joint. After the release of the fuselage latch, the pilot pushed the control stick – thus completely releasing the aft pyramid and flapping it backwards; then the pilot pulled the control stick, with the forward latches disengaging from the shaped lock – thus the aircraft separated from the TB-1's wing.

To increase the 'Zveno Aircraft' flight range, additional fuel tanks (450-liter x 2ea, 140-liter x 2ea and extra tanks 65-liter x 2ea) were installed in the bomb bay of the Tupolev TB-1. These were interconnected with the rest of the fuel reserve into a common line. From this common system, fuel lines were routed to supply the fighters. As the fighters consumed fuel from their own tanks, the mothership's flight technician used the pump to replenish the tanks of the I-4s. Additionally, each of the I-4s was fitted with two additional 12.5-liter oil tanks and a fuel gauge.

Along with the above modifications, the empty weight of the Tupolev TB-1 totaled 4,602kg. In comparison to the standard TB-1s, the take-off weight increased by 177kg. The increase in the weight – due to the installation of the fighter attachment pyramids, additional tanks and landing gear reinforcements – totaled 245kg; however, the designers managed to reduce the value by 68kg by means of removing DER-9 cluster-type bomb holders. The empty weight of each Tupolev I-4 was reduced by 10kg and totaled 965kg (the removed lower wings provided for a reduction of 20kg; however, the value was lessened by 10kg because additional oil tanks and attachment fittings were installed). Normal take-off weight of a SZ-1 constituted 10,500kg (the TB-1 – including the crew and the gasoline reserve of 1,625kg weighed 7,453kg + two I-4s with a take-off weight of 1,518kg). With a maximum fuel reserve of 2,625kg and 1,000kg of bombs, the maximum take-off weight of the SZ-1 totaled 12,500kg.

After all the implemented modifications, 'Zveno Aircraft 1' was accepted for State tests, which started on 19 May 1932. During May (20, 21, 22, 29 and 31) and June (5, 14 and 15), nine test flights and check flights involving in-flight fighters' separation were performed. An en route Monino-Kiev-Monino flight, with a gross weight of 12,500kg, was scheduled for 10 June. According to the plan, the Tupolev TB-1 was to reach Kiev, release the fighters, drop 1,000kg of bombs and return to Monino; yet because of a strong headwind and a shortage of fuel, the fighters had to separate near Bryansk. This, however, had no adverse effect on positive comments relating to the performed tests. At the final stage of the tests, on 14 June, only one of the fighters was separated. The TB-1 mothership, with another I-4 attached to it, remained stable and continued the straight flight.†‡

---

* P.M. Stefanovsky, *The Unknown 300* (Moscow, 1968).

† Zveno-1 test report, as approved on 25 June 1932: a copy from the author's archive.

‡ Archive of M.E. Zhukovsky Memorial Museum: a collection of documents relating to V.S. Vakhmistrov; files 5091/10, 5091/11, 5091/55 and 5196/134.

The tests demonstrated operational reliability of all mechanisms; there were no failures or malfunctions throughout the tests. The SZ-1 test report stated that the positive result of the creation of the 'Zveno Aircraft' comprised the guarding action of fighters during operations of the heavy bomber in the enemy's deep rear. Mastering the piloting of the SZ-1 did not require any extra skills on the pilots' part: the fighters' surplus thrust facilitated take-off of the heavily-loaded bomber from an airfield. In comparison with an ordinary Tupolev TB-1, the SZ-1 system had considerable advantages: with the same load, the service ceiling increased by 500 to 600 meters, whereas the take-off run decreased by 150 to 200m. Creation of the SZ-1 was deemed a significant achievement; it was suggested that a squadron of such aircraft should be manufactured in order to evaluate their tactical and operational peculiarities. Further suggestions included the mounting of two Polikarpov I-5 fighters on the Tupolev TB-1, and proceeding with testing the possibilities of fuel transfer from the Tupolev TB-1 and TB-3 to the Polikarpov I-4 and I-5.

A year later, the Tupolev TB-1 s/n 678 and the Polikarpov I-4 s/n 1533 and s/n 1583 were also combined into a 'Zveno Aircraft'. In this configuration, the small lower wings were not removed from the fighters, and the pilots themselves opened the release latch to become airborne. The TB-1's main landing gear was modified to contain an additional shock strut, which was designed similarly to the TB-3's 'crutch'. Operational trials of the modernized 'Zveno Aircraft' were conducted at Krechevitsy airfield in June and July 1933. The crew of the TB-1 was as follows: pilot-in-command Stefanovsky, co-pilot Vakhmistrov, installation engineer Morozov and flight technician Nikitin. The left-hand fighter was piloted by Kokkinaki, and the right-hand fighter by Stepanchonok. Again, the tests were met with approval; it was confirmed that the implemented improvements had facilitated the fighters' separation technique. Take-off from the airfield was performed at a speed of 130km/h, with a gross weight of 9,600kg, and took 12 seconds. The combined aircraft was stable in flight and allowed turns through up to 60°. In the five flights from Krechevitsy airfield, two new pilots for the I-4s and three pilots-in-command for the TB-1 were trained.

On 5 July 1933, *NII VVS* Air Brigade Commander Zalevsky signed the opinion – stating that 'The Z-1 aircraft with the above modifications is quite fit for operation in combat maneuvering units of the Workers' and Peasants' Red Army's Air Force'. Most probably, it is the 1933 tests to which another of Stefanovsky's recollections refers:

> In the course of modifications, I happened to be appointed in command of the 'Vakhmistrov's Circus' – that was how all our pilots, after somebody's joke, nicknamed the 'Zveno Aircraft'. This nickname contained both the admiration for the "triune vehicle" devised by Vladimir Vakhmistrov and a considerable grain of humor. The flight of the 'Zveno Aircraft' indeed resembled a kind of a circus show. The Tupolev TB-1 lost none of its good qualities because of the additional equipment on its wing surfaces, and kept performing all kinds of evolutions in flight, down to the deepest turns. When freed from the fighters, it behaved like an ordinary mass-produced bomber. There was, though, some occasions of circus performance indeed. I remember one of those.
>
> We were doing an ordinary mission, working on the technique of separating the fighters from the bomber. I was piloting the Tupolev TB-1. Mikhail Alexandrovich Nyukhtikov was the copilot. The fighters were piloted by V.K. Kokkinaki and I.F. Grodz.
>
> So we flew. When we entered the target area, I commanded to the 'riders' to disengage. Grodz's aircraft balanced on its forward attachments. But why is Kokkinaki shaking in the cockpit and shifting the elevator and the rudder like mad? Trying to puzzle out his 'manipulations', I forgot about Grodz. He kept balancing, waiting for the command to separate.
>
> Soon it became clear to me what was going on with Kokkinaki: Upon command to disengage, he had pulled the release lever – with the bear's might of his - and had torn off the linkage of the tail latch. And now the pilot was trying to create the needed load factors using the elevator and the rudder in order to release the latch holding the fighter's tail – to separate from us, at whatever price.
>
> But – no luck. The fighter kept seated on our wing as if it were embedded. The catastrophe

> seemed inevitable. The flight engineer (it's a pity that I do not remember his name), quite heroically, suggested that he should climb onto the wing, crawl up to the fighter's tail, and release the latch. I said NO: the airstream would have thrown him away from the wing at once. Copilot Misha Nyukhticov, my fellow student at Kachinskaya flying school, suggested shooting from his Nagant revolver to break the *so-and-so* latch with a bullet. A kind of a sniper he is! I said NO: might have shot Kokkinaki as well.
>
> Meanwhile, our unit kept flying. Grodz, without the latest command and with no idea of what had happened, continued balancing. If we allowed him off – so much worse it would be: One could not land safely with just one fighter atop the wing. And it would be much worse than just an ordinary breakage.
>
> My brain was throbbing, nerves on the stretch. What shall we do? There was still enough fuel, though. I should keep calm, should think everything over again. We do have some time yet.
>
> And here it came, the saving thought. Indeed, nobody had ever made such a decision – but there had not been such a precedent. And nobody had ever been in such a situation.
>
> The thought became converted into a final solution, "We will be landing together, all three of us", I told the crew, "Everybody take post!"
>
> Now Iustin Feliksovich Grodz sorted out what had happened as well. I signaled to him, "Whatever might happen, hold on; we will be landing together". He nodded OK in reply.
>
> We commenced descending. Everything went all right. The one thing that worried me was whether the landing gear of the TB-1 would withstand the landing load.
>
> The touchdown was smooth, and the aircraft ran along the runway. The landing gear turned out to be strong enough. But suddenly there followed a thump and cracking sound. Could it be that the aircraft was falling apart? It was not the case! The aircraft had already lost speed and stopped. What I felt then was acute worry for Iustin; I was anxious that his plane might have fallen off the wing plane. I looked in his direction and became stricken by the picture: The fighter aircraft was hanging attached to the aft pyramid frame, swinging from side to side. Up in the cockpit, the pilot was swinging like a pendulum, with the same amplitude. The Vakhmistrov's Circus indeed, wasn't it?
>
> First of all, we freed Grodz. We kept hugging him and shaking his hands. He was the one who had been exposed to the greatest risk. And it might have been worse. The matter was that during the landing run, when the rudder became inefficient because of a lack of speed, the fighter aircraft, with its tail unattached, had started sliding down from the bomber's wing plane. The aft attachment pyramid had bent through an angle of thirty degrees, ruptured the I-4's fuselage, and run into the back of the pilot's seat. The fighter's nose had pitched up badly. If the landing bump had been stronger, Grodz's aircraft would have slid off the wing plane.*

Upon completion of military tests of the SZ-1, Air Force Commander Y. Alksnis sent a memo report to the deputy chairman of the Revolutionary Military Council of the USSR Mikhail Tukhachevsky. The report stated that the payload of the Tupolev TB-1 could be increased from 3,500kg to 4,650kg – provided that it would be accompanied by the fighters both towards the target and back – and to 5,230kg if accompanied by the fighters for a distance of 50km. In the former configuration, the range of the TB-1 with one ton of bombs grew from 820km to 1,580km, whereas in the latter configuration (when accompanied to a distance of 50km), the range increased to 2,000km – thus the Supreme Command of the army and the air force displayed a favorable attitude to the 'Zveno Aircraft'. There were suggestions to award the Order of Lenin to Vakhmistrov and the Order of the Red Star to A.I. Zalevsky, A.F. Anisimov and V.P. Chkalov. In addition, it was planned to award a Ford car to Vakhmistrov and Zalevsky, and motorcycles to Anisimov, Chkalov, Panyushkin (the mechanic) and Morozov (the engineer). Vakhmistrov later wrote: 'For those flights, the

* P.M. Stefanovsky, *The Unknown 300* (Moscow, 1968).

three pilots and I were awarded a Red Star each [i.e. each was awarded the Order of the Red Star]'.*

Tests of the 'Zveno Aircraft' in the configuration of 'one Tupolev TB-1 + two Tupolev I-4' were not the only tests in 1933. Two months after completion of the tests (in September), another 'Zveno Aircraft' (Zveno-1a, or Z-1a) was configured: it consisted of a Tupolev TB-1 and two Polikarpov I-5s; military tests of the Z-1a were successfully conducted in September 1933. Air Force Commander Y. Alksnis approved the Z-1a test report on 5 October 1933 with a positive opinion and the suggestion that a series of such combined aircraft should be constructed.

Actually, in late 1933, Vakhmistrov's team within the 1st Heavy Bomber Air Brigade started work on the preparation of six TB-1 platforms for the mounting of twelve I-5s. The work for the creation of the Z-1a in such a configuration was partially completed in May 1934, with military tests to follow; however, no documents which would evidence such tests are available. Apparently, the combat unit had enough problems to deal with in the sphere of military training, and lacked enthusiasm for any risky flights in extreme conditions. Meanwhile, the more powerful platform Tupolev TB-3 came into existence to supersede the TB-1 – therefore, further tests and investigations were conducted using the four-engined TB-3 bomber.

* Archive of M.E. Zhukovsky Memorial Museum: a collection of documents relating to V.S. Vakhmistrov; a record of V.S. Vakhmistrov's interview with M.S. Arlazorov on 7 February 1969.

A test configuration with the I-4 fighter (s/n 1502) mounted on the wing of a two-engined TB-1. This initial version of the 'Zveno Aircraft' had recesses in the upper skin of the bomber to accommodate the fighter's wheels, and the pyramid-shaped frames for the fighter's landing gear attachment were of a smaller height.

The landing gear of the I-4, as attached to pyramids of initial design.

A general view of modified pyramids for attachment of the I-4 fighter on the TB-1's right half-wing. In this version, the fighter's wheels did not contact the bomber's wing; the aft pyramid for attachment of the fuselage tail portion is shown reclined.

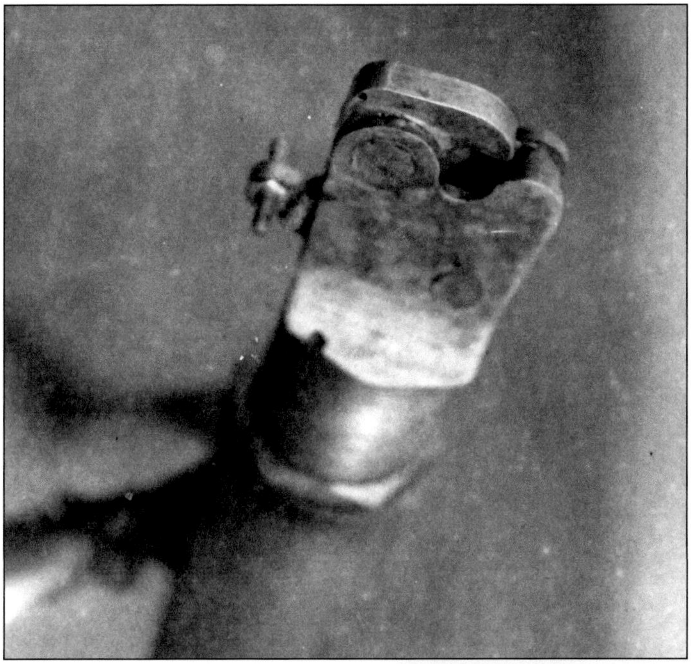

The release latch installed on the aft 'reclined' pyramid.

The strut under the fuselage of the I-4, with a special fork for engagement with the latch on the aft 'reclined' pyramid.

A fragment of the I-4 landing gear, with a visible attachment fitting and a pipeline for fuel supply from the main fuel tanks of the Tupolev TB-1.

The reinforced landing gear of the Tupolev TB-1, with an additional shock strut from the Tupolev ANT-9 aircraft.

A view of the three aircraft constituting the SZ-1 configuration at summer parking.

The power plant of the I-4 fighter, as modified for the 'Zveno Aircraft' configuration. An additional oil tank can be seen under the open door of the engine hood.

The process of pushing the I-4s up the wing of the TB-1 turned out to be far from easy.

Another photo highlighting the difficulty of pushing the I-4s up the wing of the TB-1.

The left-hand fighter mounted and secured on the TB-1 wing.

Preparation for the first tests of the 'Zveno Aircraft' on 3 December 1931 at the 17th Air Brigade's airfield in Monino.

'Zveno-1' ready to start, summer 1932.

Prior to separation of the I-4 fighter, with ski-equipped landing gear.

The moment of tail lock release: the aft truss reclined, the I-4 to lower its tail and to separate from the mothership.

A spectacular start! Two fighters separated almost simultaneously; to avoid collision in the air, the pilots of the I-4s take different directions at once.

An escort I-4 fighter accompanies the TB-1 bomber from which it was launched in flight.

The 'Zveno-1' configuration during tests in June 1932.

Participants of the 'Zveno Aircraft' tests: A.F. Anisimov, V.S. Vakhmistrov, A.I. Zalevsky, V.P. Chkalov and A.R. Sharapov.

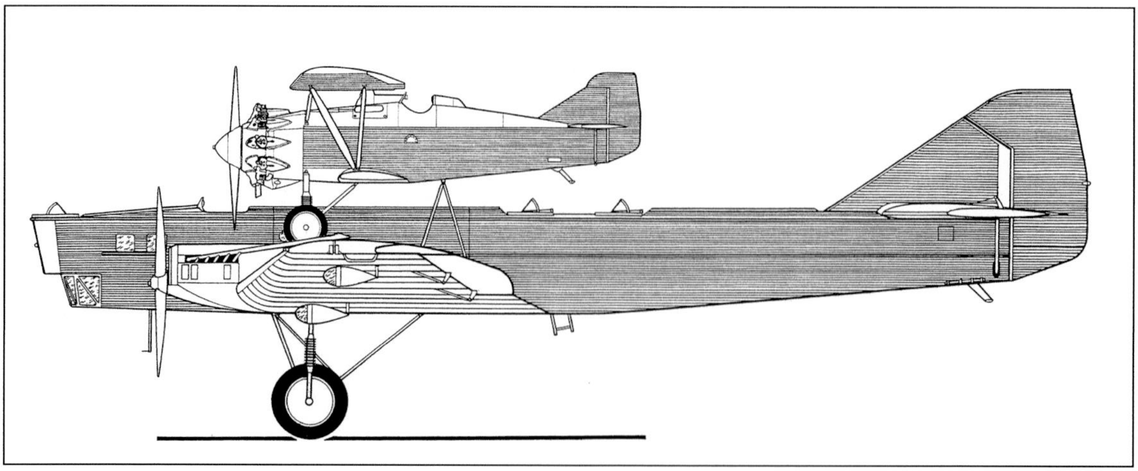

A side view of the Z-1 system (one Tupolev TB-1 bomber + two Tupolev I-4 fighters).

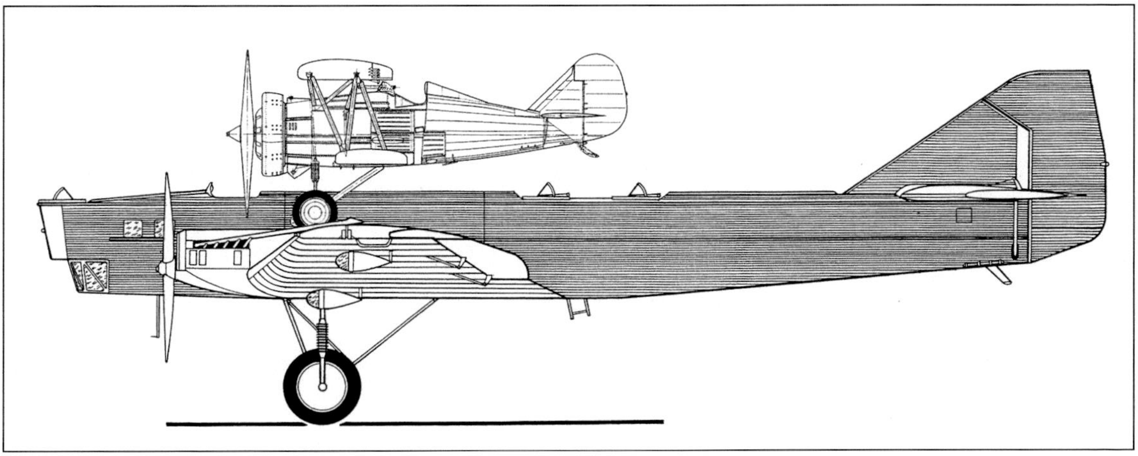

A side view of the Z-1a system (one Tupolev TB-1 bomber + two Polikarpov I-5 fighters).

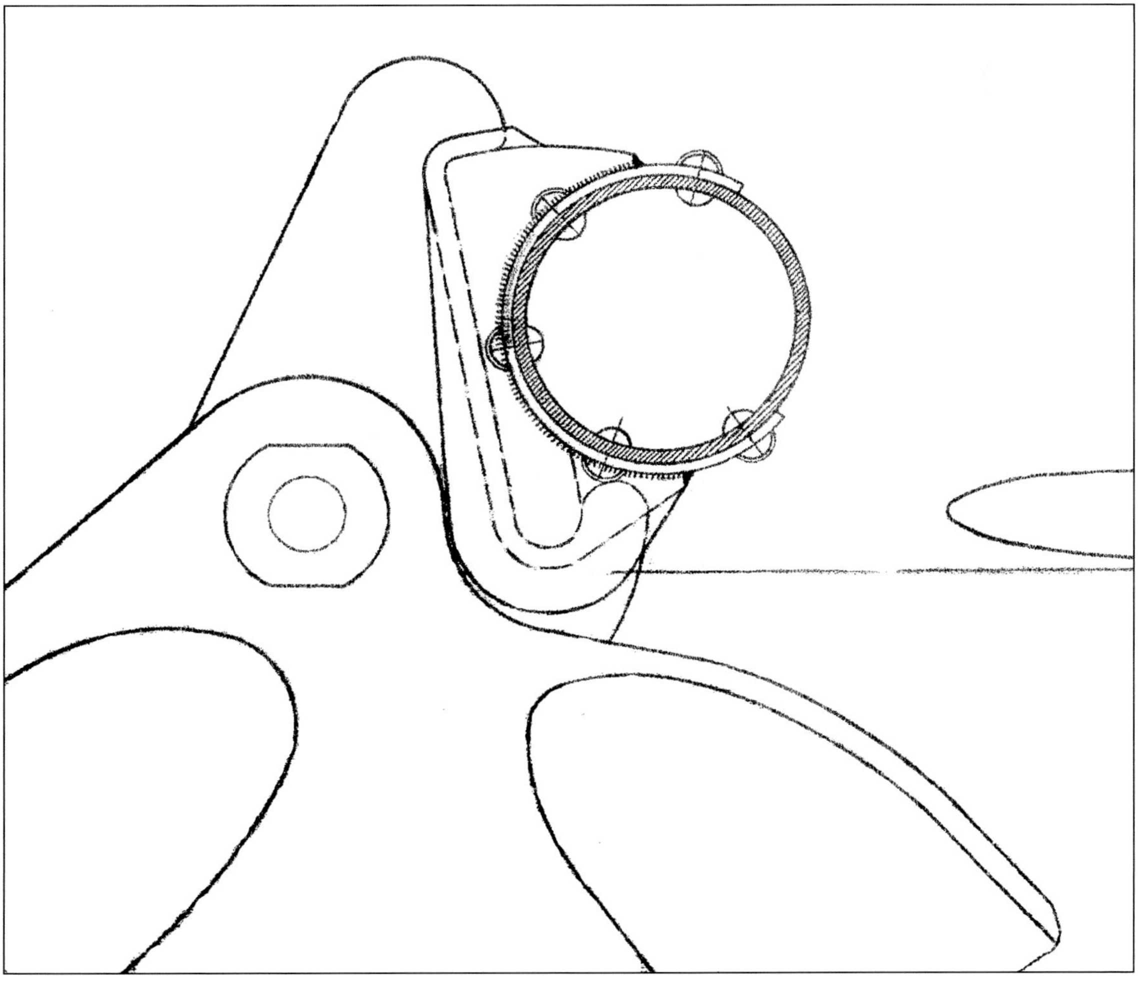

A fragment of the original drawing showing the attachment of the fighter's landing gear to the bomber's pyramid, with a special shaped support.

2

Work Continued: 'Zveno Aircraft Z-2' (One Tupolev TB-3 Bomber + Three Polikarpov I-5 Fighters) and 'Zveno Aircraft Z-3' (One Tupolev TB-3 Bomber and Two Grigorovich I-Zet Fighters)

Four-engined Tupolev TB-3 (ANT-6) bombers became operational within the Red Army's Air Force in 1932. During the winter of 1932/1933, one such aircraft was modified into a platform for another design by Vakhmistrov. Initially, pyramid-shaped frames for the attachment of two Polikarpov I-5 fighters were installed on the bomber's wing. The structure was, in general, similar to the system which was already tested on the TB-1; however, ramps for pushing the aircraft up the wing became significantly longer, and the process of rolling up the aircraft became more difficult. Sometime later, an additional Polikarpov I-5 was mounted on top of the TB-3's fuselage on supports of a special design. Most probably, this additional fighter was lifted using a crane; anyhow, the process of mounting the fighter on top of the bomber's fuselage was far from easy.

In summer 1933, the new air system – entitled 'Zveno Aircraft Z-2' and comprising a Tupolev TB-3 bomber and three Polikarpov I-5 fighters – was generally ready. Flight tests were conducted in June through to September 1933, with the mothership carrying one, two and finally three fighters. It turned out though that the design with fighters mounted on top of the wing had only been good for the TB-1. The process of pushing the fighters up the ramp in order to be mounted on the TB-3's wing was significantly more difficult, and this made the designer look for different 'Zveno' configurations. That was the reason why, in 1934, Vakhmistrov designed 'Zveno-3' (Z-3), which consisted of a four-engined TB-3 bomber and two Grigorovich I-Zet fighters (the latter were not fitted with Kurchevsky recoilless guns). While on the ground, the fighters taxied in place under the bomber's wing on the left and on the right, and were attached via a special floating frame which permitted lowering them by 0.5m before separation in flight. The first Z-3 test flight took place in July 1934, with P.M. Stefanovsky as captain for the TB-3 and A.V. Korotkov and I.F. Grodz piloting the two I-Zets. During the unlocking maneuver, the airflow lifted Korotkov's I-Zet up; the fighter crashed the skin of the mothership's wing and got stuck in it. As a result, the bomber had to perform an emergency landing. Unfortunately, the fighter (which had been stuck in the TB-3's wing) fell away and nosed over – killing the pilot. After the crash, the 'Zveno-3' project was terminated entirely.

The failure of the new configuration impeded further work, and there were no more flights in the summer of1934. As for the Z-2, despite the criticism in respect of this system, it was tested in 1935 to a full and reasonable extent.

During 15 May-13 August, tactical tests were performed for the 'TB-3 4M-17 + two Polikarpov I-5' configuration, with the following test objectives declared:

1. To check the possibility of fighters refueling from the TB-3 mothership, and to select the refueling patterns
2. To check the possibility of solving the tactical task of escorting the TB-3 over a radius of action twice the radius of action of an I-5
3. To check the possibility of repelling a sudden enemy fighter attack

The initial gross weight of the TB-3 mothership, with an aircrew of eight persons, was 19,000kg – including a bombload of 1,000kg and a fuel weight of 4,700kg. The gross weight of the entire Z-2 'Zveno Aircraft' with two Polikarpov I-5 fighters mounted (1,300kg each) was 21,600kg. During one of the final test flights, the 'Zveno Aircraft' flew along the Shchelkovo–Vladimir–Kovrov route and back. The cruising speed during the flight was 165km/h at a cruising altitude of 1,200 meters (the aircraft was unable to climb higher than 1,500m). The three aircraft flew as a 'Zveno' configuration for a distance of 570km; then the Polikarpov I-5 fighters – fully fueled – were launched from the mothership's wing, performed a five-minute demonstration of simulated air fighting and returned to the base at Shchelkovo. The total distance flown by the fighters constituted 990km. The TB-3 flew 190km more after the separation of the fighters, released the bombs within the designated training area and returned to the base as well; its flight range constituted 1,360km. The test report stated once again that the concept of a 'Zveno Aircraft' was quite viable: in particular, the tests had confirmed the possibility of carrying the fighters to a distance twice their radius of action. The range of the entire 'Zveno' configuration was, at least, equal to that of a single TB-3 bomber. At the same time, the report admitted that the airborne configuration as it was during the tests was obsolete. Flight speeds of the TB-3 bomber and the I-5 fighters were low and failed to meet new requirements – and there was no communication between the bomber and the fighters. Pushing the fighters up to the bomber's wing was cumbersome: it took at least 12 persons and six hours to carry out pre-flight procedures.

To conclude the above opinion, it was suggested that a 'Zveno Aircraft' should be created using more modern aircraft types. The report insisted that such aircraft should be built as soon as possible, with tests to be conducted by the *NII VVS*.*

* Archive of M.E. Zhukovsky Memorial Museum: a collection of documents relating to V.S. Vakhmistrov; a report on tactical tests of 'Zveno Aircraft' 'TB-3 4M-17 + two I-5s', file 5091/14.

One of the forward pyramids on the TB-3's wing to attach the landing gear of an I-5 fighter.

Pyramids to attach an I-5 fighter on the left wing of the TB-3 bomber.

Pyramids to attach an I-5 fighter: the aft pyramid for fuselage attachment is shown reclined.

The aft pyramid for fuselage attachment, shown reclined.

A Polikarpov I-5 fighter mounted on top of the fuselage of a Tupolev TB-3 bomber. Archive documents provide no idea of how it was lifted and mounted to the TB-3 fuselage. Meanwhile, that was a major problem – and it is obvious that the test participants successfully found a way out.

Wooden ramps are visible in these photos (behind the TB-3's wing). These ramps were used for the rolling-up to a height of 5m and the further mounting of two more I-5 fighters.

Zveno-2 – configuration: TB-3 4M-17 + two I-5s.

Zveno-2 – full configuration: TB-3 4M-17 + three I-5s.

A Polikarpov I-5 fighter on the right wing of the Tupolev TB-3 bomber, waiting for the air launch.

A Polikarpov I-5 fighter on the left wing of the Tupolev TB-3 bomber has released the aft lock and is ready to separate from the mothership.

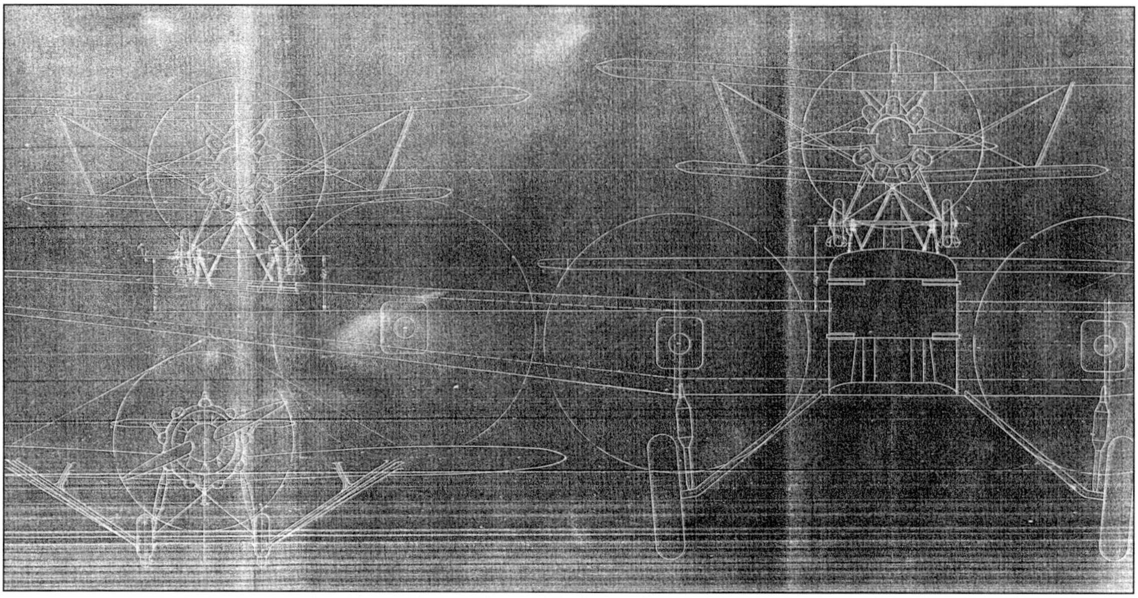

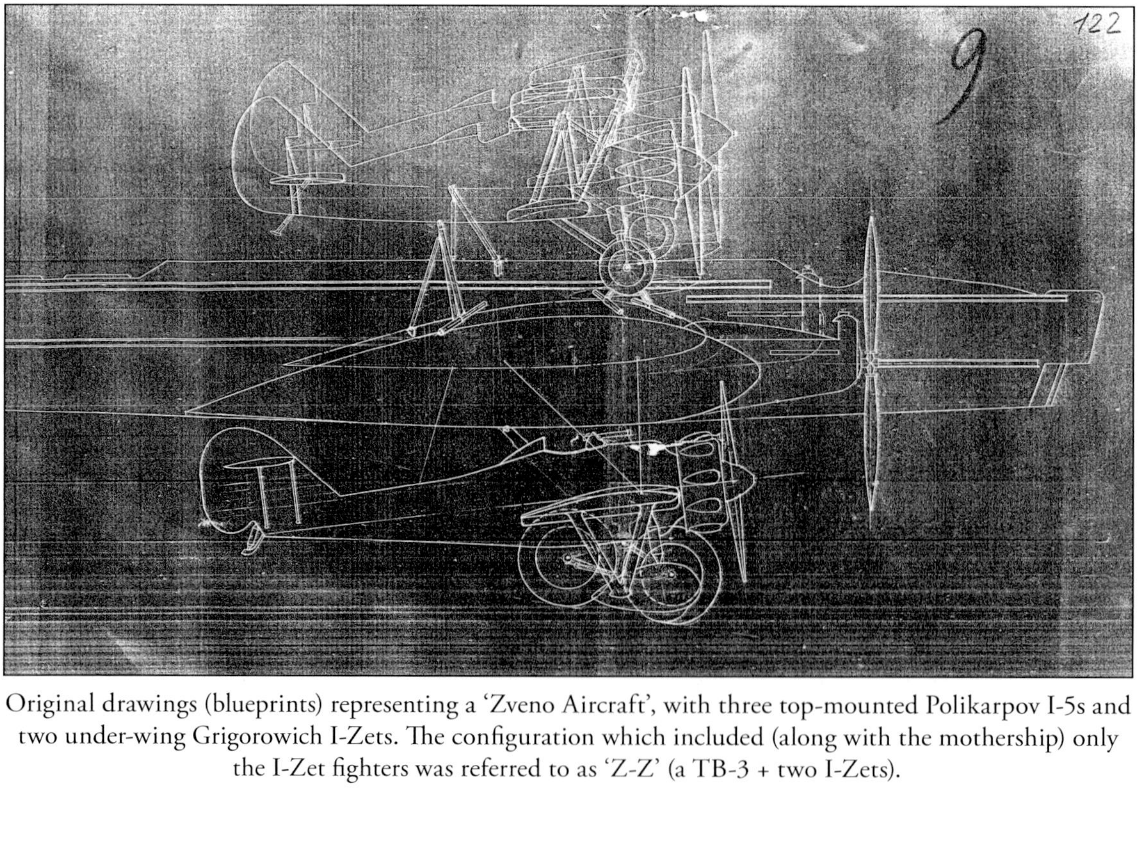

Original drawings (blueprints) representing a 'Zveno Aircraft', with three top-mounted Polikarpov I-5s and two under-wing Grigorowich I-Zets. The configuration which included (along with the mothership) only the I-Zet fighters was referred to as 'Z-Z' (a TB-3 + two I-Zets).

3

Air Attachments: 'Zveno Aircraft Z-5' (a TB-3 + an I-Zet)

After the unsuccessful tests of the configuration – with the under-wing attachment of two Grigorovich I-Zet fighters – the designer suggested attaching a single I-Zet under the TB-3 in flight. According to other sources, the development of such an attachment was planned in as early as 1933; however, the most efficient work began as soon as Vakhmistrov's team obtained an approval for the task by an Order for *NII VVS* dated 5 February 1934. In the institute, the team enjoyed the rights of a scientific sector – and according to the designer himself, the team was referred to (in 1934) as 'Vakhmistrov's Team No.5', which consisted of 10 specialists and was based at Aviation Factory No.22.

It should be noted that the idea to attach and detach aircraft in the air underwent several modifications. Initially, there was a version which consisted of the landing of an I-5 fighter on top of the TB-3 bomber fuselage. It was supposed that the fighter – fitted with a special hook – would approach the carrier aircraft from above, make its speed equal to that of the bomber and engage the hook with a light ramp (to be raised by the bomber's crew). After the engagement, the fighter would be pulled down, mounted and attached to special cradles, as arranged on the bomber's fuselage. Nonetheless, it very soon became clear that such 'air docking' between the two aircraft would be especially difficult and dangerous. In 1934, the team started work on another version, with the fighter to approach the bomber from below. This very version – involving the I-Zet monoplane – was the one to reach the design stage after many consultations with experienced pilots.

To receive the fighter, a special lowering space truss was arranged under the fuselage of the four-engined carrier aircraft; the truss was attached to aft locks of DER-15 outer bomb racks. The truss' lowering and lifting was controlled from the flight engineer's cabin using a hand-operated self-sustaining worm-gear winch. The Grigorovich I-Zet fighters were also subjected to an additional upgrade: an attachment truss – with a special hook – was arranged on top of the fighter's fuselage, and the hook was used to engage with the lowering truss from the TB-3. The hook was also provided with a retaining lock and a limiting bar to protect the I-Zet's propeller from damage. The fighter's wing structure was reinforced near ribs 12-14, because in the said area, the wing – with the fighter in a pulled-up position – was contacting the TB-3's wheels. The top portion of the I-Zet's fuselage – including the pilot's headrest – served as an additional attachment point after the docking with the bomber (therefore, this portion was also reinforced).

Tests of the configuration, which was designated 'Zveno Aircraft Z-5', started on 15 March 1935. The I-Zet fighter was piloted by Stepanchonok, while Stefanovsky was at the carrier aircraft's controls. First, special training was conducted: Stepanchonok approached the TB-3 (which was flying at a speed of 140km/h) from behind, equalized the speeds and then – opening the throttle little by little – he hooked and tore a transverse percale band, which was tied across the trapeze instead of the cross-bar. Twenty-five air training sessions of this kind were completed before 21 March 1935.

The world's first mid-air docking between the two aircraft took place on 23 March 1935. After hooking the cross-bar of the trapeze lowered from under the TB-3's fuselage, the I-Zet fighter was pulled up and secured on latches; then the Zveno-5 lowered to 500m and flew above the airfield, where an air force commission was present. The separation of the I-Zet occurred at an altitude of 1,200m. During the subsequent days – up to 29 March – five more successful air dockings were performed by Vasily Andreyevich Stepanchonok.

The following opinion was expressed in respect of the results of those flights:

1. The performed tests demonstrated an absolute possibility for the I-Zet aircraft to be attached to and detached from the TB-3 aircraft in the air. Presence of minor oscillations in the horizontal plane does not affect the attaching/detaching process.
2. The I-Zet aircraft, when attached and pulled up, demonstrates no tendencies towards falling away or oscillation about all the three axes at speeds of 125 thru 155km/h. No vibration was detected. In other flight conditions, tests were not performed due to inability to tightly press the I-Zet's wing to the landing gear wheels of the TB-3.
3. The I-Zet aircraft, when attached and pulled up, allows flying without pushing or pulling the control stick.
4. Visual communication is provided in flight during the docking; besides, communication using written messages can be implemented.
5. In such configuration, bombs may be attached to the TB-3 only under DER-13 or DER-23 cantilever bomb racks.

The Z-5 system shall be deemed successfully tested, with upgrades to be continued…*

Further, the Z-5 was upgraded as part of a 'Zveno Aircraft' comprising six aircraft.

* A report on experimental tests of an I-Zet aircraft mounted on the TB-3 aircraft: a collection of documents relating to V.S. Vakhmistrov; files 5091/13, 5091/68, 5091/56, 5091/66 and 5091/67.

One of the initial projects for the mid-air docking of the TB-3 and I-5 using a lifting ramp; this first version (aircraft landing on aircraft) was never implemented.

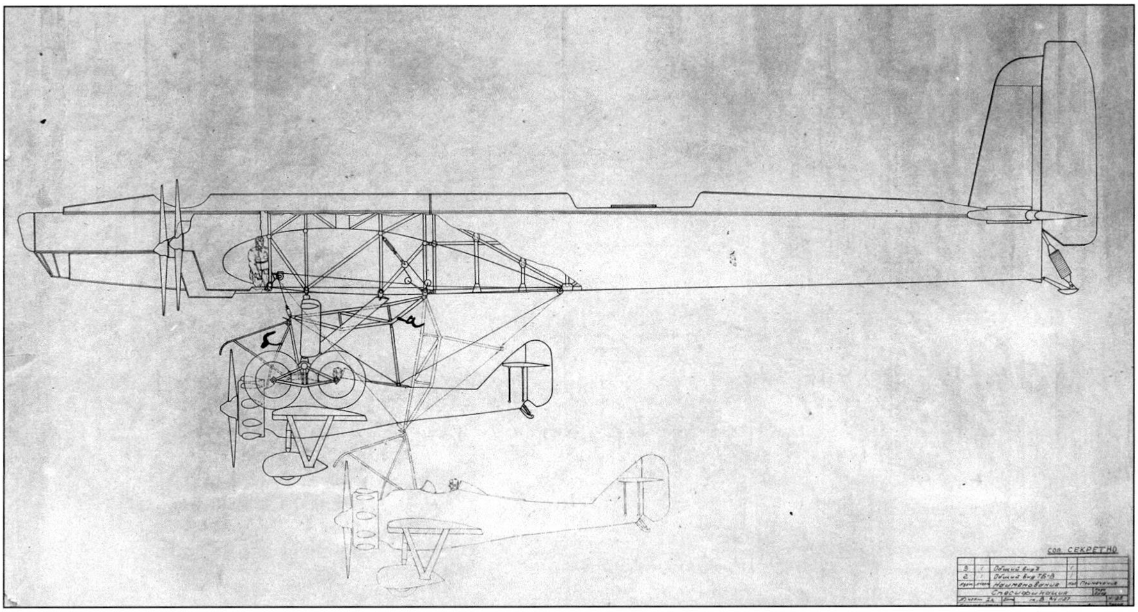

An original schematic representing the attachment of the I-Zet fighter to the TB-3 bomber.

The forward portion of the I-Zet's fuselage, with a special device for mid-air docking.

The lowering truss under the TB-3's fuselage was used to attach the I-Zet. During ground tests of the truss' lowering/lifting, the airfield personnel had to dig out a ditch in the ground.

23 March 1935: The I-Zet is approaching the TB-3 to carry out docking; this photo was taken from the R-5 reconnaissance aircraft.

The I-Zet approaching the bottom of the TB-3; this photo was taken from the TB-3's bomb hatch.

The already attached I-Zet is being pulled up by the winch to tightly secure it under the bomber.

The fighter has just got attached to the carrier aircraft for the first time.

The I-Zet secured under the bomber.

The moment of detachment.

Test pilot V.A. Stepanchonok; this photo was taken during the 1930s.

A general view of a Grigorovich I-Zet fighter armed with Kurchevsky recoilless guns.

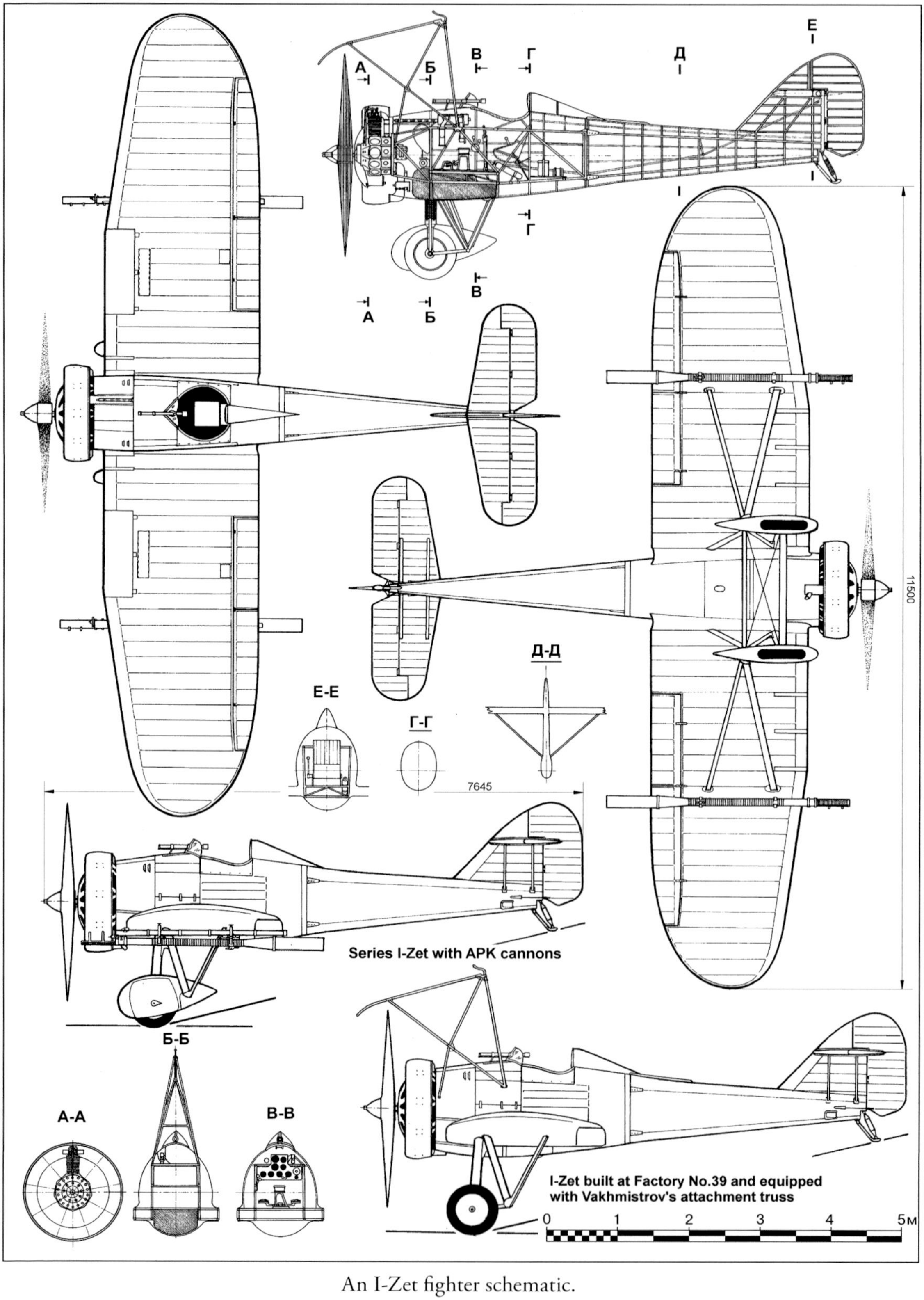

An I-Zet fighter schematic.

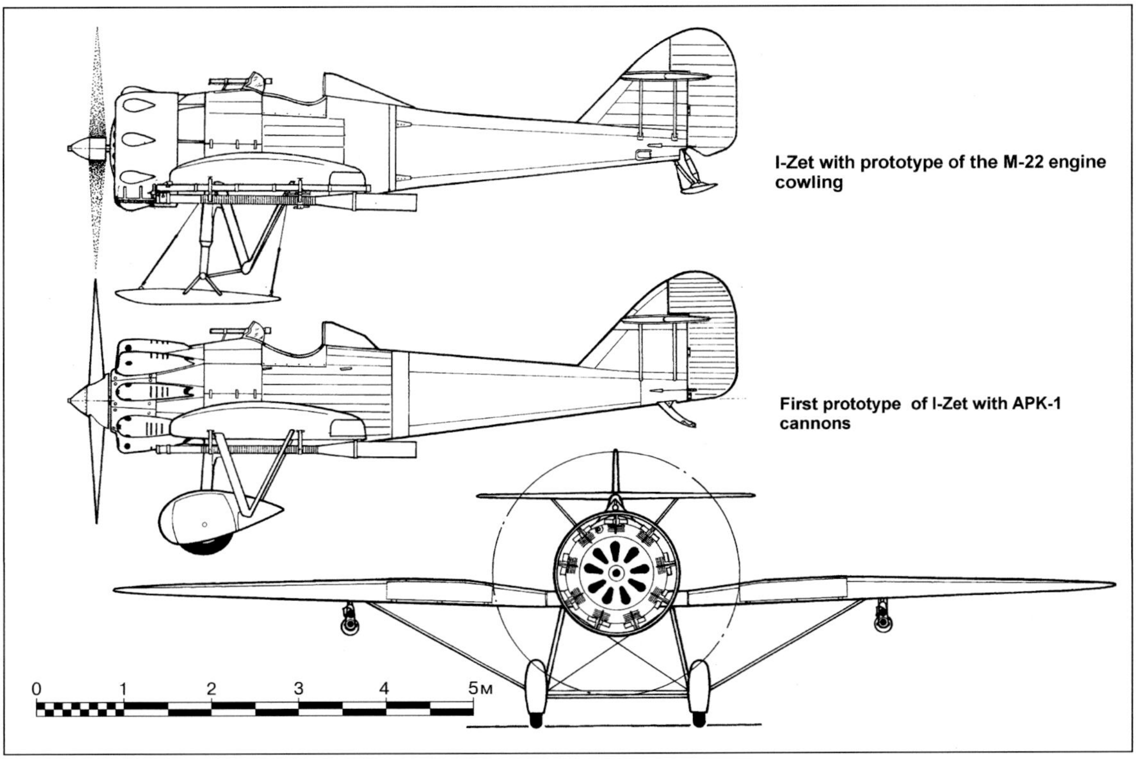

An I-Zet fighter schematic.

4

'Aviamatka PVO' (Airborne Mothership (A-M) for Air Defence) (TB-3 4M-17 + two I-5s + two I-16s + an I-Zet)

Before the autumn of 1935, Vakhmistrov prepared (for tests) another airborne system, which was designated as 'Zveno TB-3 4M-17 + 2 I-5s + 2 I-16s + I-Zet'. The novelty comprised the use of Polikarpov I-16 fighters, and this attachment version will be discussed in more detail below, but first let us note that the I-16s appeared in the Zveno configuration in a rather quick and unexpected manner – and Vakhmistrov himself referred to the combination of six aircraft as 'Aviamatka PVO' (A-M). On the ground, two I-16s were attached under the wing and two I-5s were pushed up to be mounted on top of the wing; the remaining 'participant' (the I-Zet fighter) joined the 'airborne stack' after take-off. During the tests, A.I. Nikashin and S.P. Suprun piloted the I-5s, T.T. Altynov and K.K. Budakov piloted the I-16s and V.A. Stepanchonok piloted the I-Zet. The TB-3 mothership was piloted by P.M. Stefanovsky.

He remembered:

> We were taking off from a narrow concrete runway, which was located close to the hangars. At take-off, I had to keep a perfect straight line. Eight engines with a long arm could have introduced a rather undesired 'correction' to the intentions of the crew and the command. To simplify the take-off, I had ordered Altynov and Budakov to open out the throttle of their engines simultaneously with me. And I had warned that as soon as I took the exact direction along the runway centerline I would nod. That would be a sign to Nikashin and Suprun to open out the throttle of *their* engines. There were a lot of spectators on the runway edge near the standing start. I began to apply power smoothly, starting the take-off run. The airfield buildings were sweeping by quicker and quicker. Suddenly the aircraft, like a horse uncurbed, turned right, and ran straight to the hangars. That was Suprun's fault: He was too quick to open out the throttle. I throttled back the portside engines almost to a minimum. Again, a surprise: The aircraft darted to the left. Now it was Nikashin's fault: He saw Suprun apply power, and pushed the throttle lever as well. That was probably because he was afraid that he had overlooked my nodding. Or maybe I had indeed nodded inadvertently at the moment of the aircraft's dashing towards the hangars. Anyway, I opened out the throttle of the portside engines abruptly, feeling that it was yet not too late to take-off. And we did take off. The take-off run had been a little longer than usual, though.
>
> Further, everything went in accordance with the flight task. At an altitude of 1,000 meters, Stepanchonok's I-Zet approached us right under the bomber's landing gear, and we had it 'moored'. The 'Aircraft Bouquet' or, what our local wisecrackers used to call it, 'The Airborne Pell-Mell',

> performed several wide and stately circles above the airfield. Then I signaled – and off were all the five fighters launched from the mothership at once.*

The events described by Stefanovsky took place at the airfield of the *NII VVS* during 15-22 November 1935. In total, three complete-configuration training flights and one demonstration/acceptance flight before the Air Force Commission were performed. The tests confirmed the possibility of flight for a 'Zveno Aircraft' comprising a TB-3 bomber and five fighters. Never before – and never afterward – was anything like this performed in the world.

The tests were reckoned as joint ones with Factory No.22 in Moscow, which had carried out modifications of the fighters and the carrier aircraft. The gross weight of the 'A-M' with four fighters totaled 20,200kg, and 21,700kg with the I-Zet attached. According to Vakhmistrov's estimate, which was confirmed by *TsAGI* experts, with certain reinforcements of the TB-3's wing and landing gear, the gross weight of the 'A-M' could have brought up to 28,000kg. Such a combined airborne vehicle was intended to be used as a cruiser-carrier, which would – at the same time – provide for the refueling of the fighters up to the moment of their separation.

The opinion summarizing the results of the tests stated that 'Zveno configuration consisting of a TB-3 4M-14 + two I-5 + two I-16s plus an I-Zet' could be efficiently operated by units of the Workers' and Peasants' Red Army. The document called for the development of military operational requirements to the airborne combination in question, and for gathering a meeting to finally determine the feasibility of its implementation. The document was approved by the head of the *NII VVS* Brigade, Commander Bazhanov (the director of Factory No.22 Margolin), and the head of Department 1 of the *NII VVS*, Petrov; however, the final opinion turned out to be somewhat different. It was admitted that the process of pushing the I-5s up the wing of the TB-3 was laborious; the attachment of the I-Zet near the bomb hatch reduced the TB-3's capabilities, whereas the attachment of the I-16s under the wing was simple and convenient. It was, therefore, the 'TB-3 + two I-16s' configuration that was subjected to further improvements.†

To summarize the episode, below are some Vakhmistrov's reminiscences of the 'airborne mothership':

> For the 'aviamatka', a huge Cheranovsky's flying wing ('parabola') was intended to be used. The 'aviamatka' was to be a bomber armed with bombs and also lifting six fighters. I was the one to present the project at a session of the Revolutionary Military Council. I brought all drawings and sketches there. I was given some 15 or 20 minutes to present the report. Semyon Mikhailovich Budyonny helped me hang out the sketches. The sketch representing the 'parabola' was not quite clear; first, Budyonny hung it upside down.
>
> The audience displayed a keen interest in the report. I was asked many questions, and answered them. The discussion took approximately two hours instead for 15-20 minutes. As a result, I was delegated to Factory No.22 to build the aircraft. Our project team consisted of young designers, students of the Moscow Aviation Institute. We prepared the design and successfully presented it to the mock-up committee. Then we started building the aircraft.
>
> Tukhachevsky was a great supporter of our work. However, when he presented our project to Stalin, the latter said, 'Comrade Tukhachevsky, you are a very good Marshal, but also a great dreamer!'
>
> The following day, my project was closed, and the uncompleted aircraft discarded.‡

---

* P.M. Stefanovsky, *The Unknown 300* (Moscow, 1968).

† Aviamatka PVO: A report on experimental tests of Zveno configuration 'TB-3 4M-17 + two I-5s + two I-16s + an I-Zet', in cooperation with Factory No.22: a collection of documents relating to V.S. Vakhmistrov; file 5091/15.

‡ Archive of M.E. Zhukovsky Memorial Museum: a collection of documents relating to V.S. Vakhmistrov; a record of V.S. Vakhmistrov's interview with M.S. Arlazorov on 7 February 1969.

According to the above extract, in 1933-1935, Factory No.22 started constructing a large 'parabola' aircraft designed by B.I. Cheranovsky; however, no documents pertaining to the beginning of any such activities have been found to date. Furthermore, there is no mentioning of any full-scale development of such a 'flying wing' even in Cheranovsky's archive. Most probably, a few specialists from Vakhmistrov's design bureau made the calculations and performed some other activities at the avant-project stage, but no documents have survived till now.

The airborne armada in full configuration.

An I-Zet fighter with a hook approaching under the attachment truss lowered from the TB-3.

The under-wing I-16s detaching from their attachment frames right after the center-located I-Zet.

An I-Zet separating from the TB-3.

The TB-3 carrier, with the four fighters, is ready to taxi to take-off position.

The left-hand I-5 is the first to separate from the carrier.

The TB-3, with the I-5s still attached to the wing.

The TB-3, with the five fighters in flight.

The I-Zet docking under the TB-3.

The I-Zet docking under the TB-3.

The I-Zet catching up with the TB-3 to perform the docking.

The airborne combination taking off.

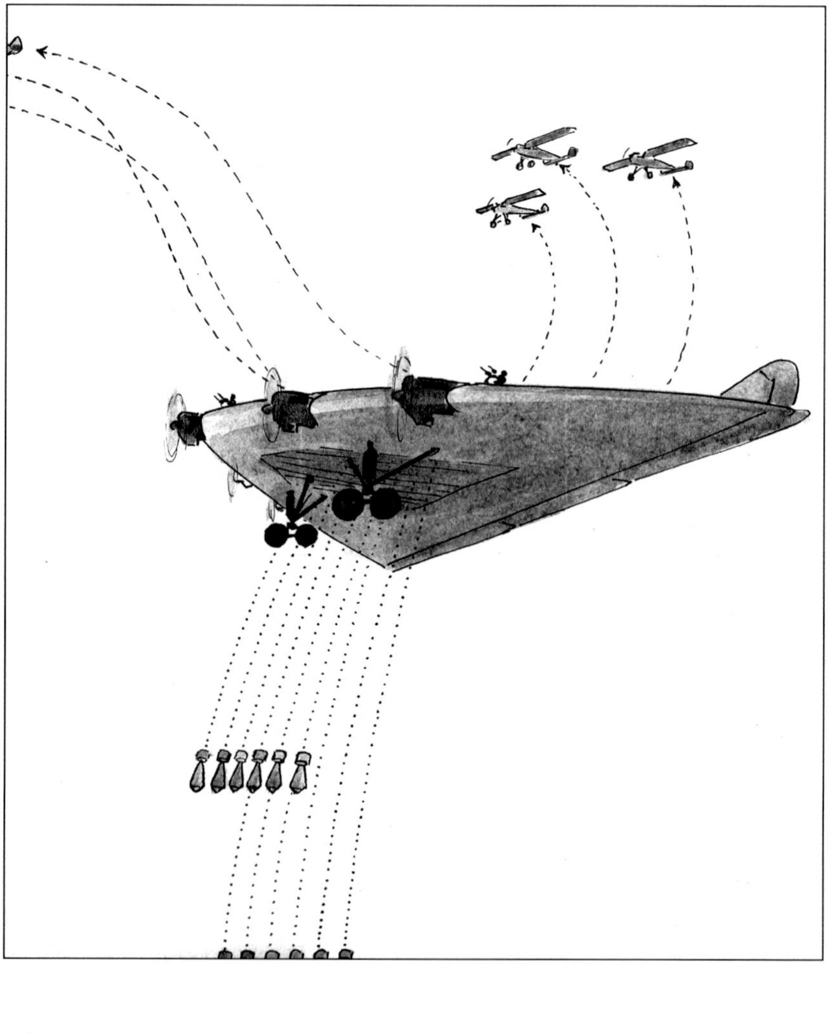

This 'Aviamatka' design version is clearly of a fantastic nature. The designer placed 15 fighters on a hypothetical flying carrier, whose contours resemble those of Cheranovsky's 'parabola'. The sketch is by the pilot Popov – one of Vakhmistrov's teammates.

5

Towing Vehicle: An Aircraft for Accelerating a Heavy Bomber-Configuration 'Zveno Z-4' (Tupolev TB-3 4M-17 + 1 M-22)

Among the special capabilities of the 'Zveno Aircraft', there was something that became clear almost immediately: a perceptible increase in the thrust-to-weight ratio of the entire combination of aircraft at take-off, which permitted to reduce considerably the take-off run and, if necessary, to increase the weight of the load to be lifted. Designer Vakhmistrov himself wrote sometime later: 'I suggested developing the 'Zveno Aircraft' concept further. Strength allowed loading the aircraft considerably, but then there was not enough power for take-off. Accelerators had not come into existence yet, and I suggested using a towing vehicle, like the one used for a glider-type take-off to lift the heavy-loaded machine from ground'.*

The suggestion to create a towing aircraft to accelerate take-off of the newest heavy bomber – the TB-3 – was presented in 1932 in the general list of Vakhmistrov's new inventions, as follows: 'Tyagach' [Russian for 'towing vehicle'] is essentially a flying engine. This is a light and inexpensive special-purpose aircraft with a powerful engine and a propeller with special characteristics. It can lift a heavy-loaded bomber and carry it to the front line'. The deputy chairman of the Revolutionary Military Council, Tukhachevsky, supported Vakhmistrov's project on 28 December 1932 and provided a favorable written opinion of the suggestion to create the 'Tyagach'.

The initial configuration – consisting in the mounting of a small monoplane on top of the TB-3 fuselage – was worked out by Vakhmistrov personally. Later, Vakhmistrov managed to engage designer P.D. Samsonov (he was a seaplane specialist in the Central Design Bureau, or *Tsentralnoye Konstruktorskoye Byuro* (*TsKB*); apparently, cooperation with Vakhmistrov was a kind of contractual activity for him), who developed and presented – in early February 1933 – a preliminary design of an elegant monoplane with a top-mounted wing. The aircraft had a full length of 8.1m and a wing span of 12m.

For preparation of the project, the designers made use of the Isotta-Fraschini Asso-1000 engine – developing the power of 1,000hp at take-off. The engine was fitted with a reduction gear, and was quite suitable to rotate the 4.5m-diameter propeller. The Asso-1000 was created in 1928 and, several years later, was still considered one of the most powerful engines in the world. With such an engine installed into a small aircraft, the latter indeed turned into a 'flying engine', with the Asso's weight of 900kg, its full length of over two meters and overall width of 1,060mm. Cooperation with Italian companies was, at the time, in full swing, so the opportunity to have the Asso engine was quite tangible; however, the exotic flying vehicle was never built. The 'Tyagach' never went further than the preliminary design stage, but still, the concept as it was

* Archive of M.E. Zhukovsky Memorial Museum: a collection of documents relating to V.S. Vakhmistrov; a record of V.S. Vakhmistrov's interview with M.S. Arlazorov on 7 February 1969.

remained of interest for a couple of years – in particular, to improve the flight performance of the TB-3 4M-17, a fuselage of an I-5 fighter was attached on top of the TB-3 fuselage, and in the Z-2 system (a TB-3 + three I-5s), one of the I-5s with its wing removed was attached over the TB-3's fuselage. The new system was designated 'Zveno Z-4' (a Tupolev TB-3 4M-17 + 1 M-22). Also, the top-mounted I-5 fuselage was referred to as a 'towing aircraft'. The weight of the towing aircraft was 975kg; the weight of the TB-3 was 17,200kg. The weight of the entire configuration totaled 18,175kg.

Pilots Stefanovsky, Nyukhtikov and Suprun made, in total, only four flights on the Z-4; two of the flights were made without starting the M-22 engine. During the tests, it was noted that the service ceiling of the TB-3 with the 'towing aircraft' increased by 350m; the rate of climb also grew (the combined aircraft was able to climb to 3,000m in seven or eight minutes). The flight speed increased by 10km/h, and the take-off run time decreased by five seconds; however, the final opinion admitted that further tests with the I-5 were not practical, since the obtained improvement of flight performance was too slight.

The final test report still suggested 'carrying out theoretical analysis of special-purpose powerful towing vehicles with various wing areas'; however, in practice, the subject was never addressed again.*

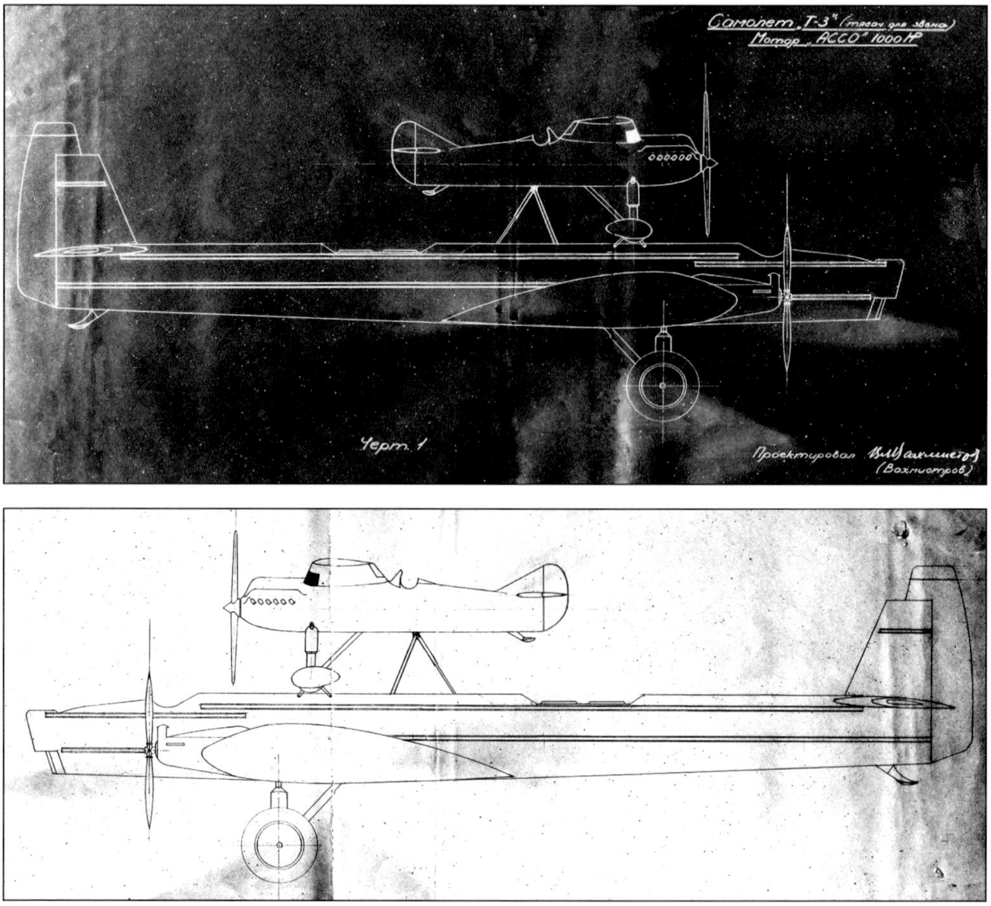

The original configuration of the 'Tyagach' aircraft, with the Asso-1000 engine mounted to the TB-3 bomber, was suggested by Vakhmistrov in 1932.

* Tyagach TB-3 4M-17+1M-22, Russian State Military Archive: collection 24708; record statement 9, file 44.

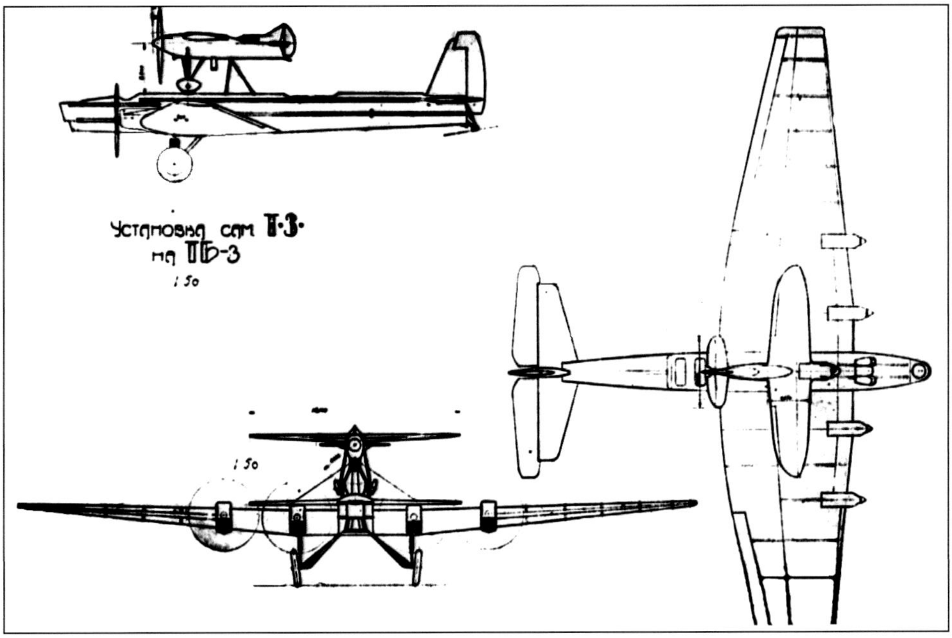

The schematic for the mounting of the 'Tyagach-Zveno' aircraft to the TB-3.

A fuselage of an I-5 fighter installed on top of the TB-3, to be used as a towing aircraft; this photo was taken in 1935.

The preliminary design of the 'Tyagach-Zveno' (T-Z) aircraft, as developed by P.D. Samsonov; the original drawing was signed by the designer in 1933.

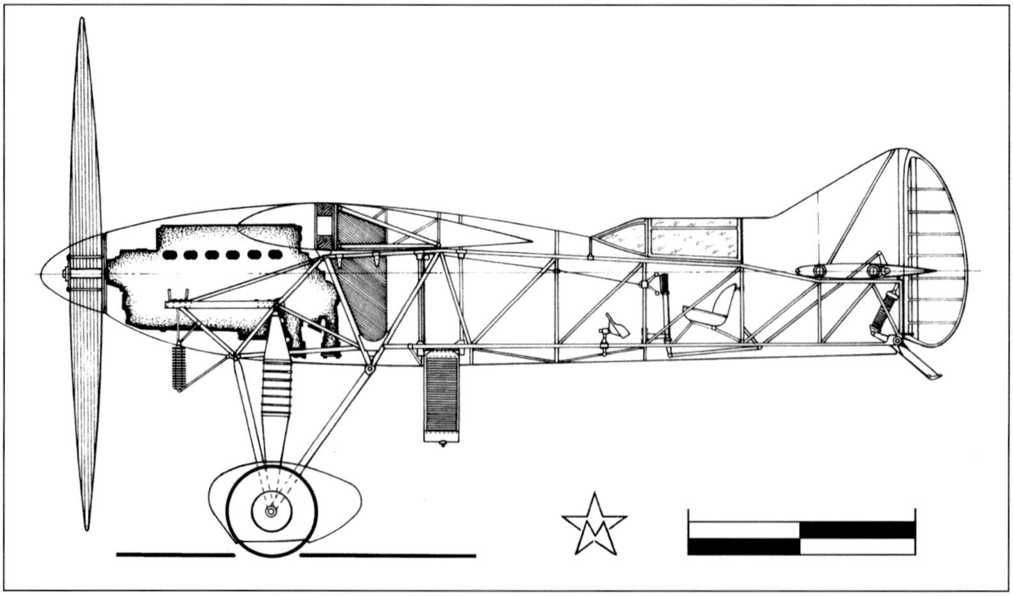

6

'Zveno Fighter Aircraft'

On 17 October 1935, the chief of staff of the Air Force Directorate of the Workers' and Peasants' Red Army – V.S. Khripin (who in November, became deputy chief of the air force) – in his letter to the Deputy People's Commissar for Defense, described all of the achievements and further plans of V.S. Vakhmistrov in detail. The final portion of the letter stated that experimental work was already finished – including the complete design (and the beginning of the construction of) a special-purpose fighter capable of taking off from and landing onto a heavy bomber:

> The work of engineer Vakhmistrov is very close to the combined aircraft which was invented in the West by Colonel Mayo …. We suggested that Vakhmistrov should make calculations of flight performance of a long-range bomber in the configuration of a combined aircraft. We have good reason to believe that such a bomber will really be of much advantage, especially if fitted with in-line inverted medium-power engines (approximately 200 to 300 HP).*

In actual practice, Vakhmistrov never dealt with combinations of heavy aircraft; at the time, he focused on creating a special-purpose small-size aircraft to be based on a heavy bomber. The designer later wrote:

> I suggested developing the 'Zveno Aircraft' concept further. Strength allowed loading the aircraft considerably, but then there was not enough power for take-off. Accelerators had not come into existence yet, and I suggested using a towing vehicle, like the one used for a glider-type take-off to lift the heavy-loaded machine from ground. At the same time, I suggested creating a super-speed fighter, which would take off from and land to an airborne mothership. Such a fighter had no limitations in terms of landing speed; thus, it could be 1.5 times faster than the aircraft existing at the time, and could develop a speed of approximately 600km/h. I almost succeeded in building the fighter. It was being built at the facilities of Factory No.22. I had a team of my own there. Tukhachevsky supported me again. His resolution on my report stated the following: 'The towing aircraft design to be developed; 2. A super-speed fighter to be designed in 1933; 3. An airborne mothership to be designed'.†

The designing of the conceptual aircraft, with a reduced wing area, was really started as early as 1933. Initially, it was a high-wing aircraft, which was intended to be attached to the TB-3 above the latter's wing using the attachment frames, which had been previously used to secure the I-5. The fighter had no landing gear; it was intended to land to an extendable ski. However, the project was soon rejected, and no further development followed.

---

* Russian State Archive of Economy: collection 8328; record statement 1, file 797.

† Archive of M.E. Zhukovsky Memorial Museum: a collection of documents relating to V.S. Vakhmistrov; a record of V.S. Vakhmistrov's interview with M.S. Arlazorov on 7 February 1969.

In 1934, in the course of developing the attachment of the Grigorovich I-Zet under the Tupolev TB-3 – the idea of creating a special-purpose aircraft referred to as 'Zveno Fighter' – underwent considerable changes, and technical requirements to such a fighter were approved on 29 August 1934 by Air Force Commander Y.I. Alksnis. The 'Zveno Fighter' was described as a high-speed machine gun fighter based on board a heavy airborne vehicle. Its intended missions included combat air patrolling while staying on the airborne mothership, airborne surveillance, escorting a group of heavy bombers and destroying the enemy's aircraft. With a payload of 300kg, the 'Zveno Fighter' had to ensure the dry-tank range of 600km, the maximum speed of 500km/h at an altitude of 4,000 meters and the ceiling of up to 10,000 meters.

The designing of the 'Zveno Fighter' – in accordance with technical requirements as provided by the *NII VVS* – started in January 1935. The aircraft was, essentially, a monoplane with an 'inverse seagull' wing; this allowed increasing the distance to the carrier mothership's wing and permitted the landing on the wing portion protruding downward in case of a forced landing. The structure was an all-metal one, with a smooth skin. The armament included four synchronous machine guns and wing machine guns, which were able to change the setting angle by 10°. For the power plant, the designers intended to use the Gnome-Rhone 14krsd (M-85) engine with a 3.4m-diameter propeller.

The aircraft was designed without any landing gear or 'crutch', since it was intended to be attached under the mothership, and landing on the ground was considered an extreme condition. According to calculations, the aircraft was able to land at a speed of approximately 130km/h using the lower surfaces of its wing, with assumed damage to the propeller and tail unit. The prototype was planned to be flight-tested with an ordinary landing gear, with wheels in fairings. In 1935, design of the aircraft was almost complete and the construction work started at Factory No.22; however, in December 1935, all work on the 'Zveno Fighter' project was abandoned by the decision of the authorities. The main reason for such a decision was the successful use of the Polikarpov I-16 fighters in the 'Zveno Aircraft' configuration. These fighters, to a large extent, were suitable for the tasks to be performed by the 'Zveno Fighter'.

Principal Technical Specifications and Flight Performance of the 'Zveno Fighter'

Wing span (m)	7.75
Length (m)	6.7
Wing area (m^2)	10
Gross weight (kg)	1,590
Load on wing (kg/m^2)	159
Maximum speed (km/h)	572
Service ceiling (m)	10,000
Time to climb to an altitude of 5,000m (minutes)	6.2 (with a 3.4m-diameter variable-pitch propeller)

Along with the above-described version, Vakhmistrov's archive contains another project of a special-purpose 'Zveno Fighter', which was designated 'Airplane D' (most probably, it was the same 'Zveno Fighter' in a new configuration). Apart from common coinciding dimension parameters, it is known that the aircraft was intended to be provided with an M-85 engine (Gnome-Rhone 14krsd) and a 3.2m-diameter propeller. There is no other information available on any design activities involving the 'Airplane D'.*

There was another interesting episode with an attachable aircraft which could have been used in Vakhmistrov's configurations. The I-17 fighter, which was designed by Nikolai Polikarpov, could have become such an aircraft. The fact was that the prototype of the I-17 fighter (referred to as 'TsKB-15') was not accepted in 1936 because of drawbacks in the landing gear system – therefore, Polikarpov prepared a project

* A collection of documents relating to V.S. Vakhmistrov.

for modification of the aircraft and designated it 'I-17 (TsKB-19) Suspended'. The project – dated 30 April 1936 – obtained no other designations. It was referred to as 'TsKB modification to be suspended under the Tupolev TB-3 heavy bomber' – hence the denomination of a 'suspended fighter'. The principal difference comprised of the reduction of the wing span to 7.6m and the reduction of the wing area from 17.65m^2 to 9.13m^2, which permitted a considerable increase in speed performance (up to 550km/h). The aircraft had no landing gear; landing on ground was considered an emergency case. The aircraft was intended to detach in the air whenever necessary and, upon completion of the mission (i.e. escorting the bombers), to dock with the mothership. According to calculations, the mid-air docking was to be carried out at a speed of 180-200km/h. The 'suspended fighter' was to be fitted with the Hispano-Suiza 12Ybrs engine with water or glycol cooling, and to be armed with two ShVAK cannons, or with a ShVAK cannon and two ShKAS machine guns. In Polikarpov's opinion, to implement the project, a new wing and new horizontal tail were required; this could have been achieved with minimum effort and time. However, the suggestion was met with no interest and was never developed in detail.*

Meanwhile, Vakhmistrov himself did not abandon the idea to design a high-speed aircraft with a reduced wing surface area. In the report prepared for the hearings in the *NII VVS* on 27 February 1938, Vakhmistrov – among other reports on combined aircraft – suggested creating a super-speed aircraft in the record-setting or fighter version. A quite unexpected suggestion was to use, as a power plant, the Renault Bengali 450hp engine. License for this French engine was purchased in 1936 – and two years later, it was being prepared for mass manufacturing under the designation 'MV-12'.

The following design parameters were specified for the suggested aircraft:

	Record-setting version	**Fighter version**
Gross weight (kg)	810	1,030
Wing area (m^2)	3.0	3.8
Load on wing (kg/m^2)	270	271
Armament	--	2 machine guns
Maximum speed (km/h)	730	650
Minimum speed (km/h)	200	200

Apparently, Vakhmistrov suggested using the French engines for the reason that the aviation industry authorities at the time were facing a kind of a 'top-priority task' to somehow make use of the Renault engine (on account of the fact that the purchased engines had not been in much demand); however, the designer's suggestions did not receive any noteworthy positive response and no other mentioning of them has survived since then – and furthermore, mass production of the MV-12 engines was never launched.

Meanwhile, Vakhmistrov kept including the concept of the super-speed fighter among his other suggestions – in particular, the letter of 27 November 1938, which was signed by the head of the Naval Aviation Directorate (Korobkov) and addressed to the head of the Design Bureau No.29 (Chizhevsky) stated a considerable interest in the work on combined aircraft (Vakhmistrov's team was, at the time, part of Design Bureau No.29); however, according to the same letter, 'a super-speed fighter, which could not land on ground, would hardly be of practical use'.

* Russian State Archive of the Moscow Area: collection 4610; file 77.

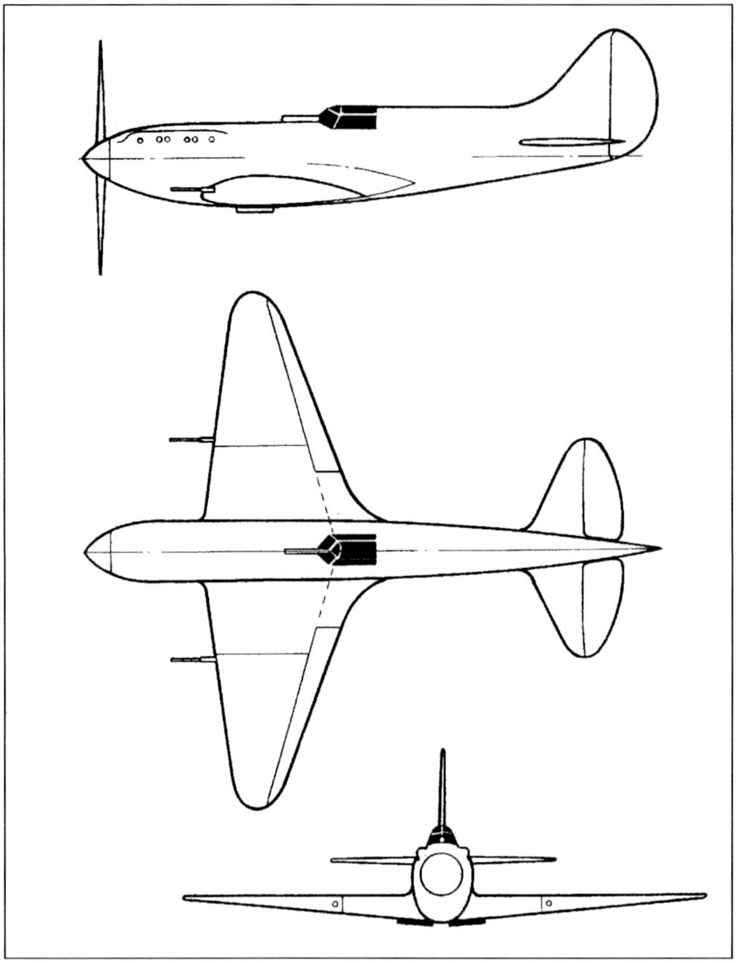

A project drawing of the suspended I-17 (TsKB-19).

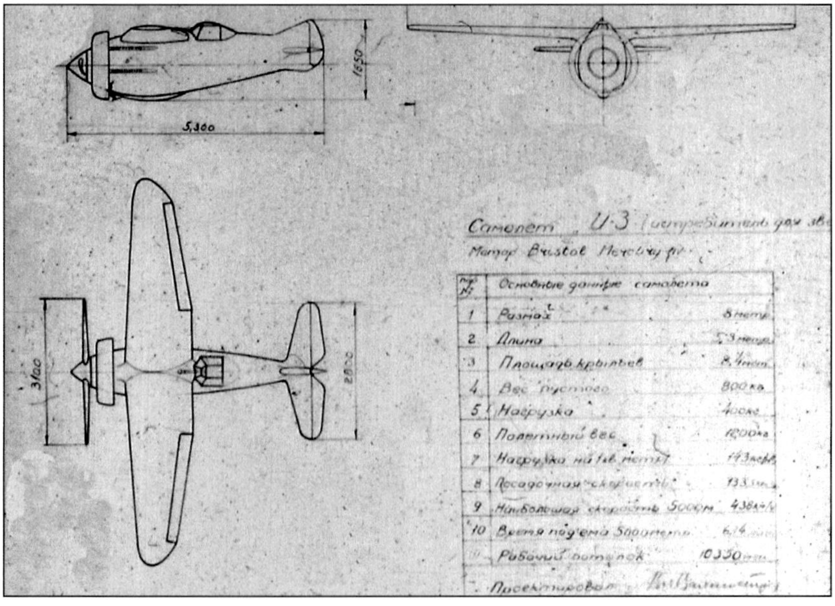

A project drawing of the 'Zveno Fighter', dated 1933.

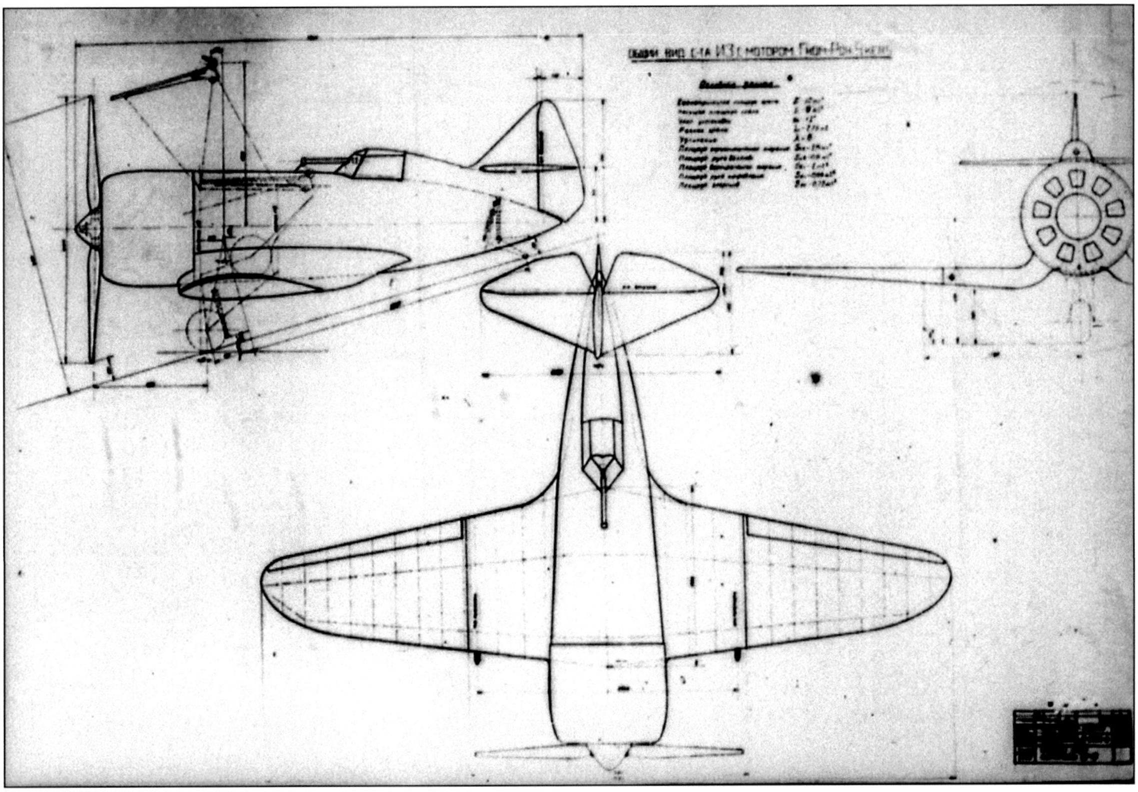

A project drawing of the 'Zveno Fighter', dated 1935.

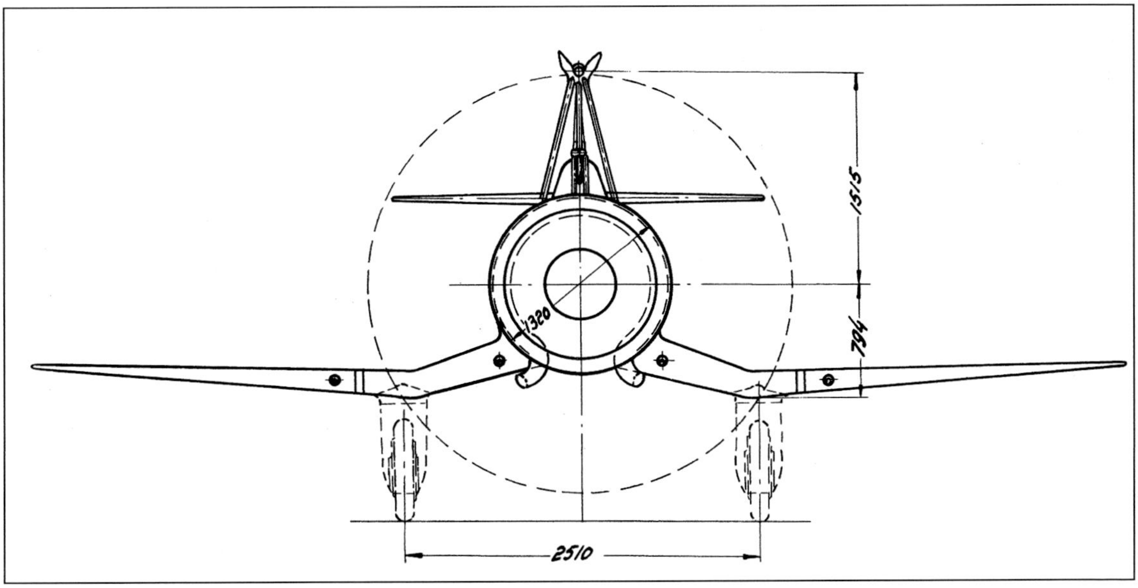

Original project drawings showing views of 'Airplane D'.

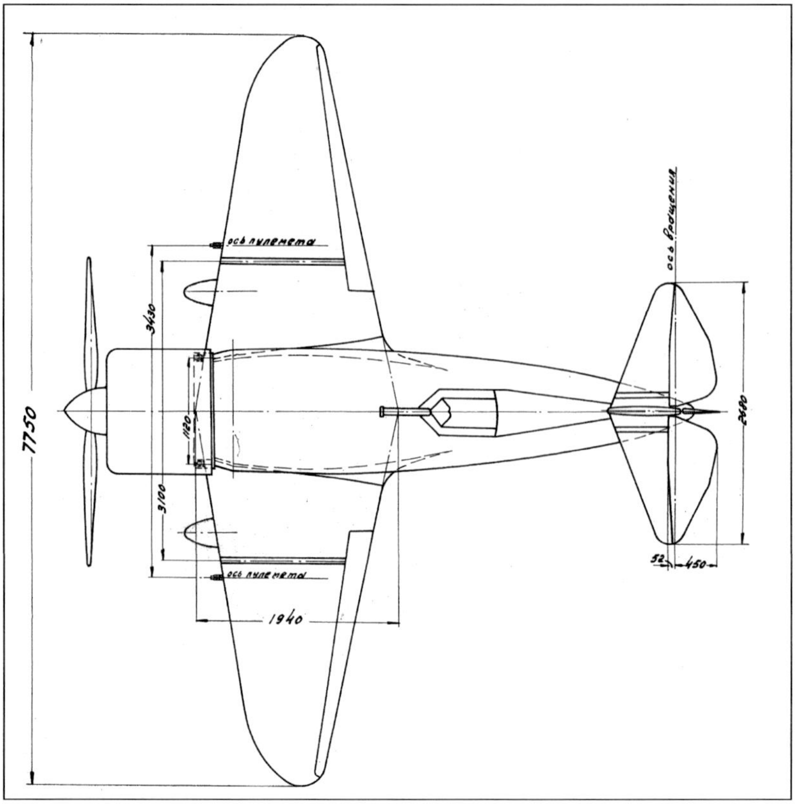

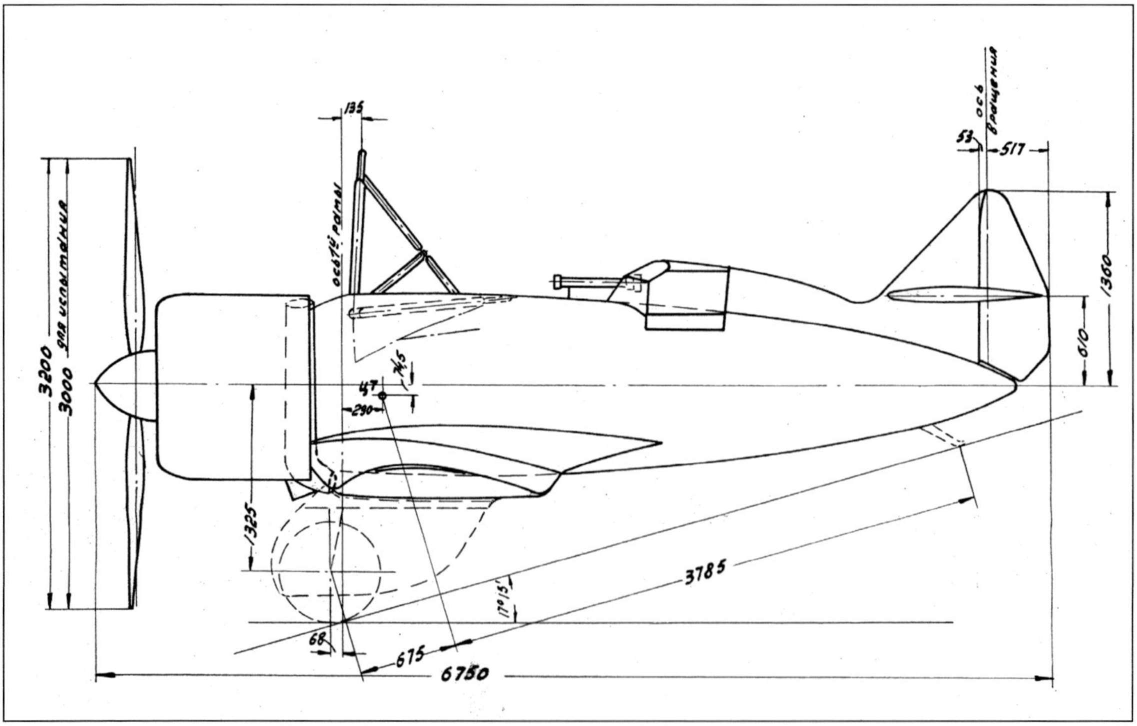

Original project drawings showing views of 'Airplane D'.

7

'Zveno-6' (TB-3 4M-17 + two I-16s with M-22 engines)

In spite of the impressive flights of the 'Aviamatka' – comprising six aircraft – this cumbersome 'flying air-stack' was not developed any further; however, the designers managed to single out from the configuration the most promising combination of aircraft – namely two I-16s attached under the wing of the TB-3. The first tests took place as early as the end of November 1935, but it became clear at once that the system required modification. Final recognition of the configuration with the two I-16s came in January 1936, in the course of preparation of a general opinion on the results of the 'Aviamatka A-M' tests. What most specialists especially liked was that the two small fighters could taxi in under the wing of the four-engined bomber, and rather quickly attach themselves to the attachment pyramids. According to optimistic estimates, the entire procedure took some ten to fifteen minutes. The four-engined TB-3 4M-17 and two I-16 M-22s were the same as those used in the 'Aviamatka' configuration; however, the attachment fitting components were subjected to considerable improvement.

Two additional load-carrying ribs were installed on each of the I-16s at the interface between the center wing section and the wing panels; the ribs were, essentially, welded trusses. Each of the trusses accommodated a robust lock to attach the wing of each fighter to the strut on the carrier aircraft. The aft (tail) lock, which was mounted on Frame 8 of the I-16's fuselage, was attached to the third strut on the TB-3. On the whole, the system was reasonable and elaborate. Each of the fighters taxied in on the port or starboard side under the TB-3, and was installed manually – and more precisely – under the attachment pyramids with the retention locks, which were 'hanging down' from the wing.

The next operation consisted in the lifting of the I-16's tail and securing the upper aft fitting. Tools and accessories used for the job included the standard canvas harnesses, which were 2.5m long and 80-100mm wide (these were provided together with each 'Burro' – a nickname for the I-16; in Russian, 'Ishachok'), and a strong trestle or an airfield trolley. After the securing of the tail portion, the nose portion of the I-16 was lifted using a BL-3 winch. The servicemen on the sides supported the wings using the belts until the wing locks were secured. The entire procedure took ten to fifteen minutes. (According to other sources, five people could complete the procedure in 30 to 35 minutes.) At this point, the attachment process was considered complete; then, a special-purpose flexible hose was attached to the fighter's port side to ensure refueling of the fighter with gasoline from the TB-3's tanks. In the cockpit on the left side, there was a special quick-release lock. With this lock released, the fuel supply from the TB-3's tanks was cut off and the I-16 automatically switched over to fueling from its own fuel tank.

Taking into account the almost twofold increase in the range, the I-16s were fitted with an additional 25-liter oil tank. At the final stage of preparation for the combined flight, an intercom system and light indication and aural warning system cables were connected. Also, a light annunciator was installed on the lower surface of the TB-3's wing in front of the I-16's pilot, so that the latter could see redundant indications of detachment.

Initially, the system was prepared for tests in January 1936; however, there is no information on any flights for the period. Apparently, the detachment tests on the ground revealed a need for additional improvements, and the flight tests were shifted to summer; however, according to the recommendations, it was winter when the leading 'Zveno-6' formation should have been tested in the *NII VVS*. If the configuration entered mass production, one of the 'Zveno Aircraft' was intended to take part in the summer air force military training exercises.

The configuration with final improvements – referred to as 'Zveno-6' – was prepared for flight tests in the middle of 1936. The declared objective of the tests was to determine the reliability of the I-16's refueling from the TB-3 in flight, and to determine the reliability of the lifting system, the intercom system, the light indications and the aural warning system. Throughout the tests, 19 attachments/detachments of the aircraft on the ground and four flights with the separation of the I-16s in mid-air were performed; in-flight separation of the fighters was carried out by the I-16s' pilots on their own. After release of the aft lock, the I-16's pilot pulled the control stick and the aircraft became separated from the forward lock. All the test separations were performed at the flight speed of 150 to 160km/h. Refueling from the TB-3 and the device for activation/deactivation of fuel supply operated normally and were recommended for mass manufacture. The light annunciators operated normally and were convenient for the pilots. The SPU-7 intercom system provided normal communication (except for maximum engine settings).

On 23 August 1936, the left I-16 spontaneously separated at take-off; this resulted in damage to the aircraft. The pilot of Aviation Factory No.22, Budakov, was not injured; the separation was reportedly brought about by a design defect in the aft lock. Additional test flights were conducted in September, and the report on the work performed was prepared in December 1936. The final report recommended manufacturing a series of equipment sets to be installed to the existing aircraft. The Red Army Air Force Commander Y.I. Alksnis suggested providing fifty Tupolev TB-3M-17s and a hundred I-16 M-22s for the creation of the 'Zveno Z-6' aircraft as soon as possible;* however, nothing was to follow at the time.

For a number of reasons, further improvements of Vakhmistrov's inventions were suspended for at least a year. The first reason was that as early as the end of 1935, Andrei Nikolayevich Tupolev was appointed Deputy People's Commissar for the Aviation Industry – in other words, Chief Engineer for the Aviation Industry. Tupolev strived to 'clean' the territory of Aviation Factory No.22 from any and all design teams which were not connected with the manufacture of A-M aircraft. In 1936, it was decided that the mass production of Tupolev TB-3 heavy bombers should be wound down, and full-scale production of high-speed SB bombers should be started. Under the circumstances, the mass-manufacture load on Aviation Factory No.22 was becoming exorbitant. The situation was further aggravated by the Resolution of the Council of Labor and Defense of 11 May 1936 on the specialization of aviation factories and on 'attaching' chief designers to mass production facilities. A.A. Arkhangelsky – appointed chief designer for Aviation Factory No.22 – viewed his principal objective to be consisting of improving, first and foremost, the SB high-speed bombers.

In view of the situation, the majority of the design teams – which were operating within the aviation industry – were converted into design offices and transferred to various aircraft-building factories. Under the circumstances, Vakhmistrov had to move to the facilities of Design Bureau No.29 in Podlipki, in the Moscow area (sometimes in business correspondence referred to as 'Mytishchi'). The Flight Test Facility (FTF) of Design Bureau No.29 was engaged in testing and improving the bombing armaments for the People's Commissariat of the Defense Industry. That was where Vakhmistrov created – and started improving – the combination of the TB-3 with the I-16 in the configuration referred to as 'Combined Dive Bomber Zveno-SPB'.

---

* A collection of documents relating to V.S. Vakhmistrov, file 5091/16: a report on additional tests of Zveno configuration 'TB-3 4M-17 + two I-16 M-22s'.

The TB-3 4M-17 with the I-16 M-22s attached under the wing. The bomber bears the traces of previous tests: attachment fittings for four more aircraft are provided on the wing and on the fuselage; this photo was taken on the airfield of the *NII VVS*, 1936.

An I-16 M-22 with its landing gear extended under the right wing of the TB-3. The I-16 belongs to one of the initial series, with its wing machine guns removed.

The I-16 attached to the pyramid has 'retracted' its landing gear and is ready to take off in the 'Zveno-6' configuration.

The BL-3 winch in the Soviet aviation industry was one of the most common devices for the hoisting of cargoes and the attachment of air-delivered bombs. It is this winch which was used to lift the I-16s and secure by the locks.

The nose portion of an I-16 is being pulled up using the BL-3 winch to mount the I-16 to the wing locks. Only the frontal section of the fighter's hood is being mounted. A flexible hose for the refueling of the I-16 during the combined flight is routed from the lower surface of the TB-3's wing to the I-16.

The hoisting sling from the winch is being routed to the hoisting point on the fighter's engine.

A view of the wing lock between the center wing section and the outer wing of the I-16.

The TB-3 left wing and pyramids for the attachment of the I-16's wing.

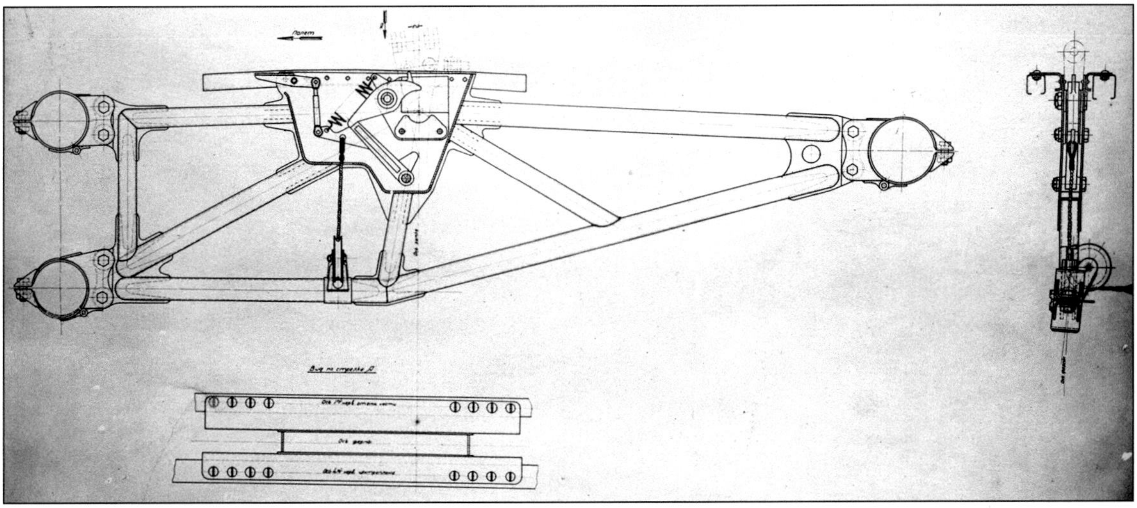

An additional load-carrying rib with an attachment lock, as installed on the I-16 in the space between the center wing section and the outer wing; what you see is a fragment of the original drawing.

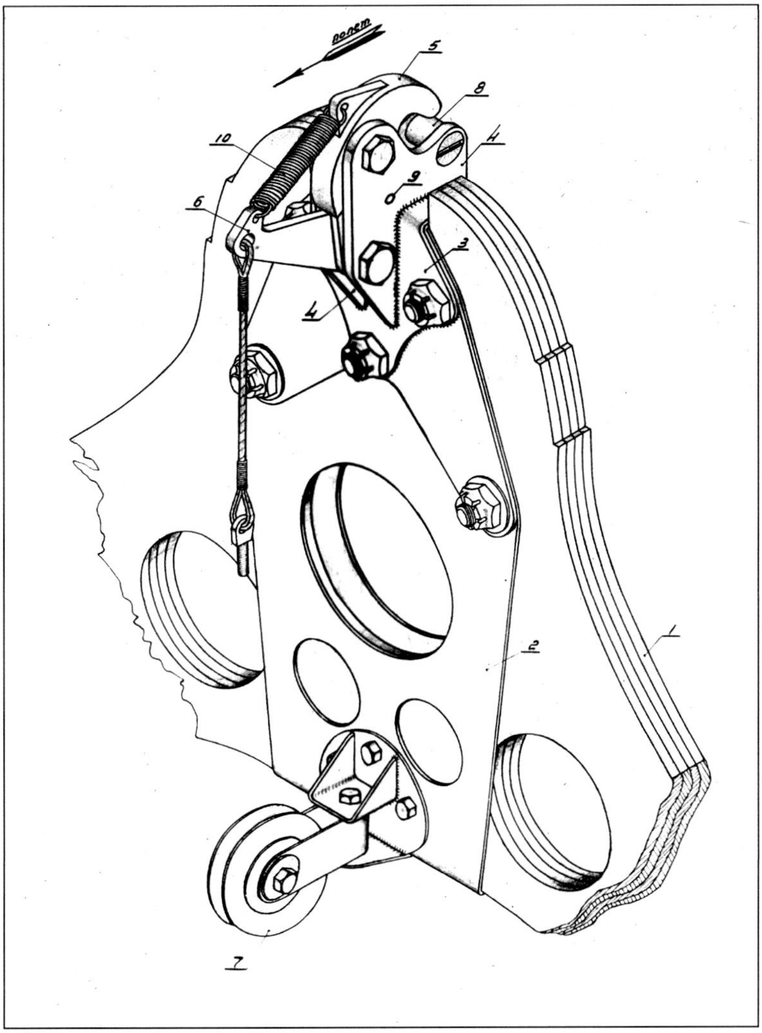

An original sketch representing the tail attachment lock on Frame 8 of the I-16's fuselage.

The tail attachment lock on the I-16's fuselage.

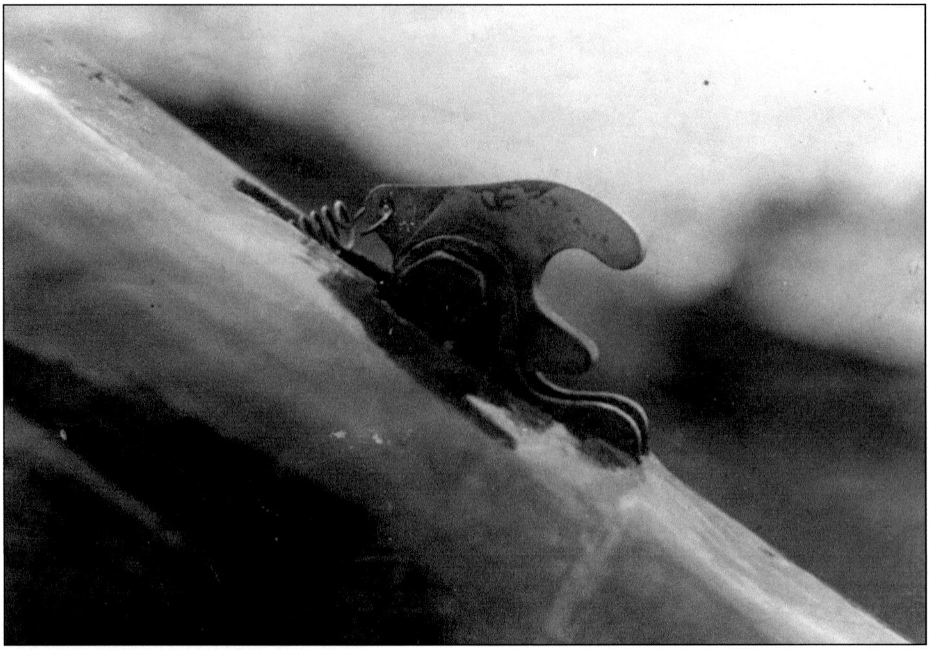

The flexible hose – via which gasoline was supplied from the TB-3's tanks to the I-16.

A special valve on the port side in the I-16's cockpit cut off the fuel supply before the separation of the fighter from the carrier aircraft.

The light annunciator under the TB-3's wing was used to provide an additional indication of the separation from the carrier aircraft.

The light indication control panel on board the TB-3 was used to provide signals to the I-16's pilots.

A fragment of the I-16's wing with the locks secured and the intercom connector, the light indication system and the aural warning system connected.

8

'Parachute-Cable Cannon'

In 1937, Vakhmistrov's inventive spirit manifested itself in a somewhat different direction than suspended aircraft: the new contemplated concept was referred to as the 'Parachute-Cable Cannon'. Though the project is far from our principal subject of the 'Zveno Aircraft', it fully complies with the general heading of the study – namely 'Vakhmistrov's Circus'. This is the reason why the author has decided to provide a brief description of the project.

The inventor himself described the 'Parachute-Cable Cannon' as a secondary weapon for fighter aircraft to repel the enemy bombers' air strikes. The 'Parachute-Cable Cannon' was, essentially, a 3kg combative charge tied to a small (1.75m) parachute by means of a 100m long cable. When stowed, this equipment occupied a cylinder 265mm long and 90mm in diameter. It was pushed to the outside by compressed air (8 to 10atm). The assembled 'Parachute-Cable Cannon', together with the ejector mechanism, weighed 18kg; of these, the weight of the charge with the cable and the parachute totaled 8kg. The intention was to use the 'cannon' in front of a formation of aircraft approaching from the opposite direction. After firing, the parachute rocketed up and opened in three seconds, with the 3kg charge hanging down on the cable. The rate of descent of the charge, as connected with the parachute by means of the 100m cable, was 6.5m/s. If an aircraft flying from the opposite direction became engaged with the cable, the parachute pulled the charge up, and – at the moment of the charge's contact with the aircraft – an explosion followed.

General opinion of the invention was quite favorable: it was believed to be efficient – and therefore, worthy of being constructed and tested. In early 1938, the 'Parachute-Cable Cannon' was mounted on a Polikarpov R-Z bomber; tests were conducted during 17 May through to 14 June 1938. A Tupolev R-6 Paravane, with special cables stretched from the wing's leading edge to the wing tips, took part in the tests; aerial observation and camera recording was performed from a Miles Hawk aircraft.

The test results provided the basis for an opinion stating that the 'Parachute-Cable Cannon' had excellent aiming performance in cases of a head-on attack; however, the 'Parachute-Cable Cannon' project was never implemented in terms of practical operational use.

A single 'Parachute-Cable Cannon' cylinder mounted on a Polikarpov R-Z bomber, with 'USSR-I189' insignia.

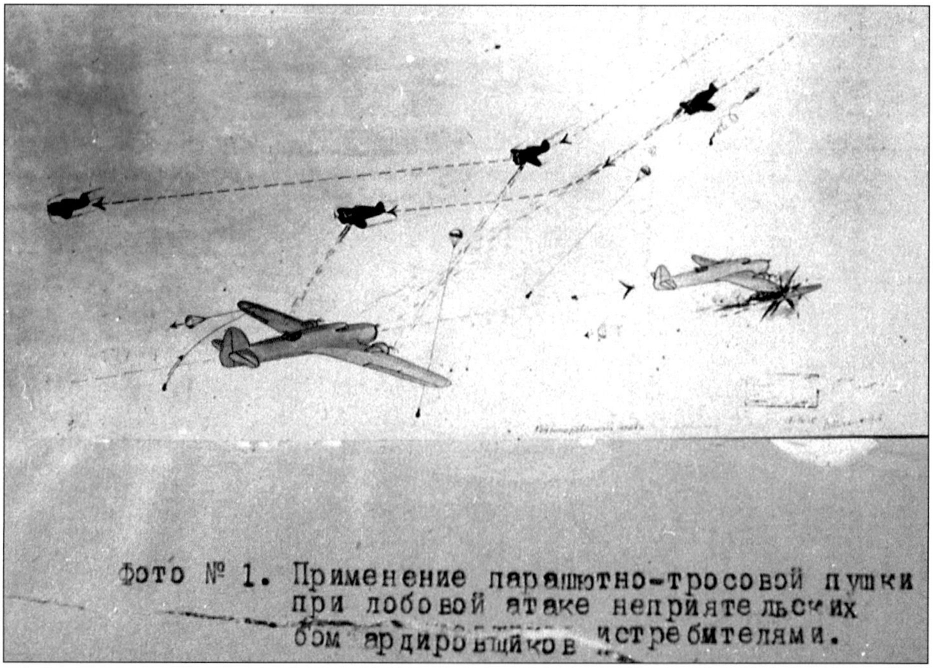

An original sketch representing the operating principle of the 'Parachute-Cable Cannon'.

9

'Zveno Aircraft SZ-7' (Tupolev TB-3 4AM-34RN + three Polikarpov I-16 M-25s)

Vakhmistrov kept improving the combination of a heavy bomber with high-speed fighters – and as early as 1937, suggested combining four aircraft (a Tupolev TB-3 4AM-34RN + three Polikarpov I-16 M-25s) into another 'Zveno' configuration (referred to as 'Z-7'). In this configuration, two fighters were arranged under the bomber's wing and the third one was attached under the bomber's fuselage. The configuration included an improved TB-3 with AM-34RN engines and I-16 Type 5s with M-25 engines. According to the designer's expectations, the new system was to offer additional advantages in terms of flight performance and tactical parameters. In this configuration, the fighters could attach to the TB-3 both on the ground and in the air. In case of the fighters' ground attachment under the TB-3's wing, the fighters' radius of action was increased twice; in case of mid-air docking of the I-16s, the fighters' radius of action was increased to an almost unlimited extent – more precisely, it was limited only by the bomber's fuel reserve, because during the flight in a combined configuration, the fighters were refueled from the TB-3's tanks. The large power margin ensured by the fighters in the Z-7 configuration contributed to the enhancement of the flight performance of the entire combination.

The Z-7 configuration could be used to move the fighters over long distances – and as an 'Aviamatka PVO', to keep a number of fighters in the air (with refueling from the mothership). All the other ideas of the inventor were also made use of within the new version. The attachment of the fighters to the TB-3's wing had been generally tried out before; in case of a mid-air docking, the fighters attached to the TB-3 as follows: a special attachment fitting ('cabane') with a lock was mounted in the top portion of the I-16 fuselage (between the engine and the cockpit); riveted duralumin beams were arranged under the TB-3's wing and fuselage. The beams were rotated – by means of cables and winches – in vertical planes parallel to the aircraft centerline. The leading ends of the beams accommodated bellcranks, which moved translationally in agreement with the rotation of the beams, so that the bars of the bellcranks were always parallel to the TB-3's centerline. For a fighter's docking under the TB-3 (under the wing), the beam rotated to retract the bellcrank and the fighter attached to the bellcrank bar by means of its 'cabane' lock; then, the I-16 was pulled up to the TB-3 and its wing was pressed to the stopping shock absorbers on the vertices of the wing pyramids. The aft point of the I-16's fuselage was attached to the aft pyramid by the respective lock. For a fighter's docking under the TB-3's fuselage, the procedure was the same as for under-wing docking, with the TB-3's landing gear tires acting as the stopping shock absorbers.

According to the initial concept, such a system for the attachment of the fighters onto the TB-3 ensured the following:

1. Take-off of the heavy bomber with two fighters mounted on it (with wheeled landing gear fully retracted or ski landing gear (in winter conditions) pulled up under the fighter's fuselage)
2. Take-off of the heavy bomber with two I-16 SPBs mounted on it (under the same conditions)

3. Flight of the combined aircraft (a Tupolev TB-3 4AM-34RN + three Polikarpov I-16 M-25s) in all normal flight conditions of the heavy bomber
4. Supply of fuel and oil to the fighters (at take-off and in flight) from the TB-3 tanks via special hoses lowered from the bomber
5. Mid-air docking of the fighters to the TB-3, with the switching to refueling from the TB-3's tanks
6. Quick detachment of the fighters in flight

In practice, the mid-air docking system was subjected to considerable and repeated modifications.

To test the concept, a so-called 'Zveno-7-1' was built; it included a TB-3 and an I-16. A special retractable truss was arranged under the TB-3's right wing – and the lowering and lifting of the truss was controlled from the TB-3's wing. After engagement of the I-16 with the truss, an operator staying in the wing pulled the fighter to the bomber's wing using a manual drive. The I-16 was provided with its own truss (or 'cabane') with a fork (yoke), which in turn, was fitted with a lock. The lock opening was controlled from the fighter's cockpit. In the pulled-up position, the I-16 was butting with its wing and fuselage tail portion against the stops, which had the shape of small pyramids fitted with shock-absorbing pads.

The impacts of the lock against the bar were absorbed by hydraulic dampers, which were included in the circuit of winch cables and the truss, and by the turning of the winch with the brake (at high impact loads). There was one-way communication by means of a signaling light/annunciator installed on the TB-3; the signaling light or annunciator was arranged under the TB-3's wing, and the signaling light/annunciator control panel was installed in the bomber's cockpit.

After the manufacturing and installation of the equipment, the TB-3 was ferried on 25 May 1938 from Podlipki to Chkalovskaya to perform tests. Preparation of the materiel took two weeks. It was noted that the TB-3-carried aircraft had already been in operation for a long time, and its cooling radiators were leaky and subjected to repair from time to time. The first training approaches – at a speed of 155-160km/h – were performed on 9 June. For the training, a 3m-long bar was arranged under the bomber's wing, with a fabric cone attached to the bar. The fighter had to approach the bomber and grip the cone using the yoke. Test pilots Nyukhtikov and Datsko took turns in flying the TB-3 as left pilots, with Antokhin as copilot. The I-16 was piloted by Suprun and Stefanovsky.

The events were filmed from a small two-seat Miles Hawk aircraft (as far as it's known, a Miles Hawk was brought from Spain) and the tests continued, with attempts to attach to an operational retractable truss under the right wing. On 3 July 1938, Stepan Suprun was the first one to attach the I-16 to the truss; during the attachment process, the 'cabane' on the I-16 was bent. On 26 July, the I-16 was piloted by Stefanovsky, who continued the tests with the training bar, but the bomber's weight turned out to be too turbulent for the small fighter; again, the 'cabane' on the I-16 was impacted and damaged. For this reason, the length of the training bar was increased to 5m. On 5 August, Stefanovsky again approached the bar and hit it with the I-16's propeller. At this point, it was decided to stop the flight tests, because the results were far from brilliant: twenty-four flights on the TB-3, 19 flights on the I-16 and only one attachment under the carrier aircraft. By the way, the report on the tests described above also mentioned the earlier factory tests – during which five attachments to the lowered bar had been performed. In fact, the I-16 had been approaching the TB-3 in a very disturbed flow; it had been constantly rolling and sucked under the TB-3, so it had been extremely difficult to engage with the carrier. The presence of the 'cabane', with the lock on the fighter, did not affect the flight performance; it was noted that its speed near ground was decreased by 8km/h. As for the 'cabane' itself, its strength and damping performance proved to be insufficient. It was suggested that the size and strength of the 'cabane's' pyramid should be increased, and the fork spread width should be enlarged to 600mm. The 3m downward extension of the lowered truss proved to be quite insufficient as well – and the influence of the bomber's wing on the attached fighter was considerable. Nevertheless, in spite of the problems and drawbacks of the new 'Zveno' configuration, all participants of the events acknowledged that the docking was quite possible.

The final document of 31 October 1938 suggested that the docking system should be improved, mounted on a more advanced TB-3 M-34RN and presented for tests in March 1939. Indeed, a new lowering mechanism for docking under the TB-3's wing was developed. For better safety, the lowered truss length was increased to 5.5m; it was now lowered in the FWD flight direction and provided with a yoke, which accommodated a shock absorber with a stroke of 250mm. Conversely, the I-16 was fitted with a small trapeze with a transverse beam; the height of the trapeze was increased by 300mm so that it could avoid the zone of the rotating propeller. The new version of the Z-7 was prepared for implementation in 1940; however, this time, the configuration never came into existence in practice.*

* A collection of documents relating to V.S. Vakhmistrov, file 5091/22: 'Zveno Aircraft' SZ-7 (TB-3 4AM-34RN + three I-16 M-25s); file 5091/17: Zveno-7 test report.

An I-16 Type 5 with a truss and lock for attachment.

A truss for the attachment of a fighter under the TB-3's right wing, in the deployed position.

A truss for the attachment of a fighter under the TB-3's right wing, in the stowed position.

An I-16 attached and pulled under the TB-3's wing; with the landing gear retracted, the fighter will be ready to take off in combination with the bomber.

An I-16 attached and pulled under the TB-3's wing; with the landing gear retracted, the fighter will be ready to take off in combination with the bomber.

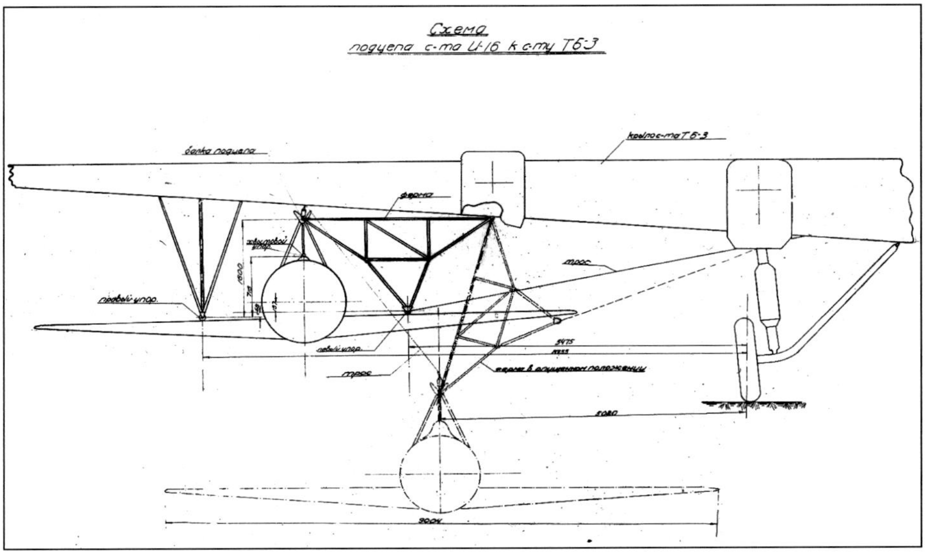

A schematic for the attachment – and further pulling-up – of a single I-16 to the TB-3's wing.

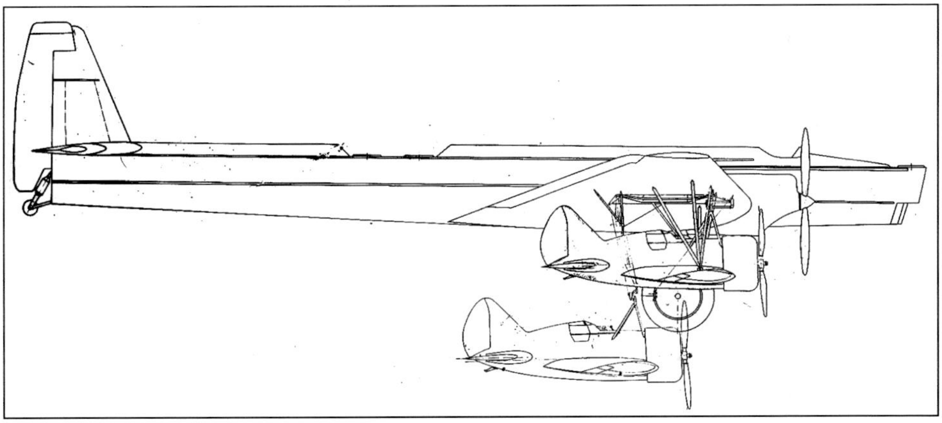

The initial version of the 'Z-7', with three fighters. The center-located I-16 is represented attached, but not yet pulled up to the landing gear wheels.

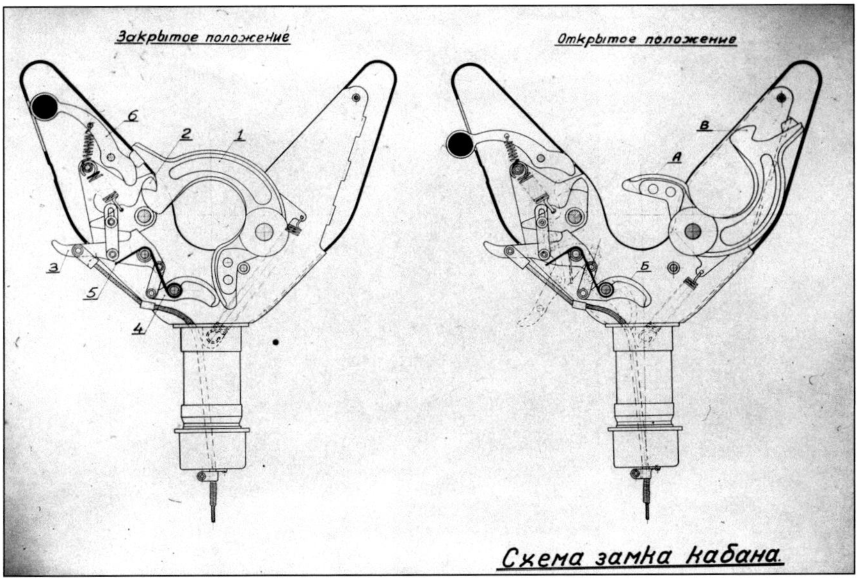

An original drawing of the I-16's fork (yoke) for attachment with the lowering truss.

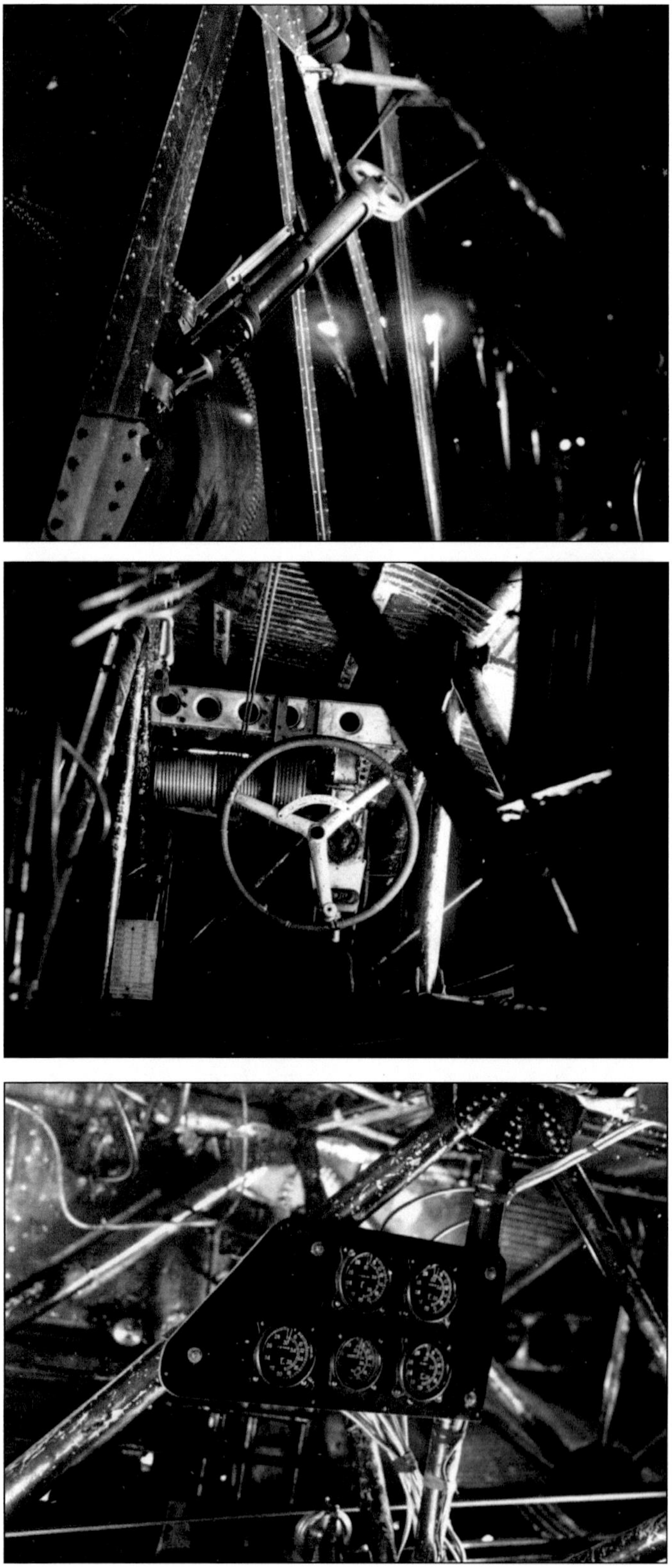

An operator's 'workstation' in the TB-3's wing to control the pulling up of the I-16 to the bomber.

Stills from a film recording an I-16's training approach to a fabric cone.

Further stills from the film recording an I-16's training approach to a fabric cone.

A final batch of stills from the film recording an I-16's training approach to a fabric cone.

Stills from a film recording S. Suprun's attempt to attach under the TB-3 on 3 July 1938.

Further stills from the film recording S. Suprun's attempt to attach under the TB-3.

An I-16 with a bent truss after an unsuccessful docking.

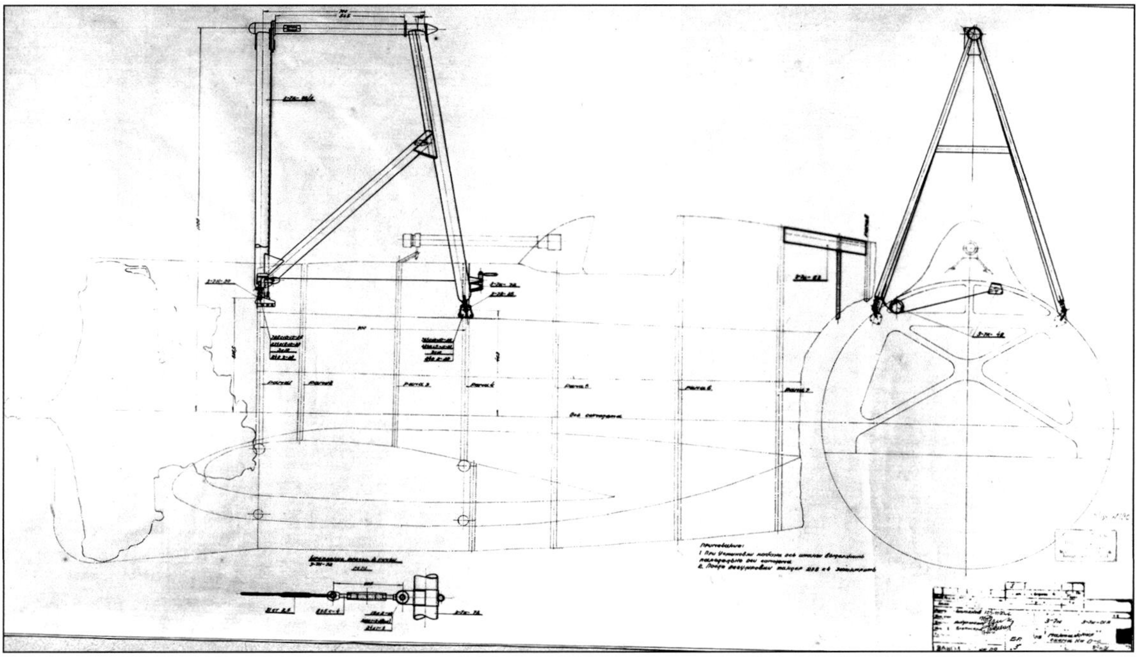

A section of the original drawing of the modified 'cabane' on the I-16, 1939.

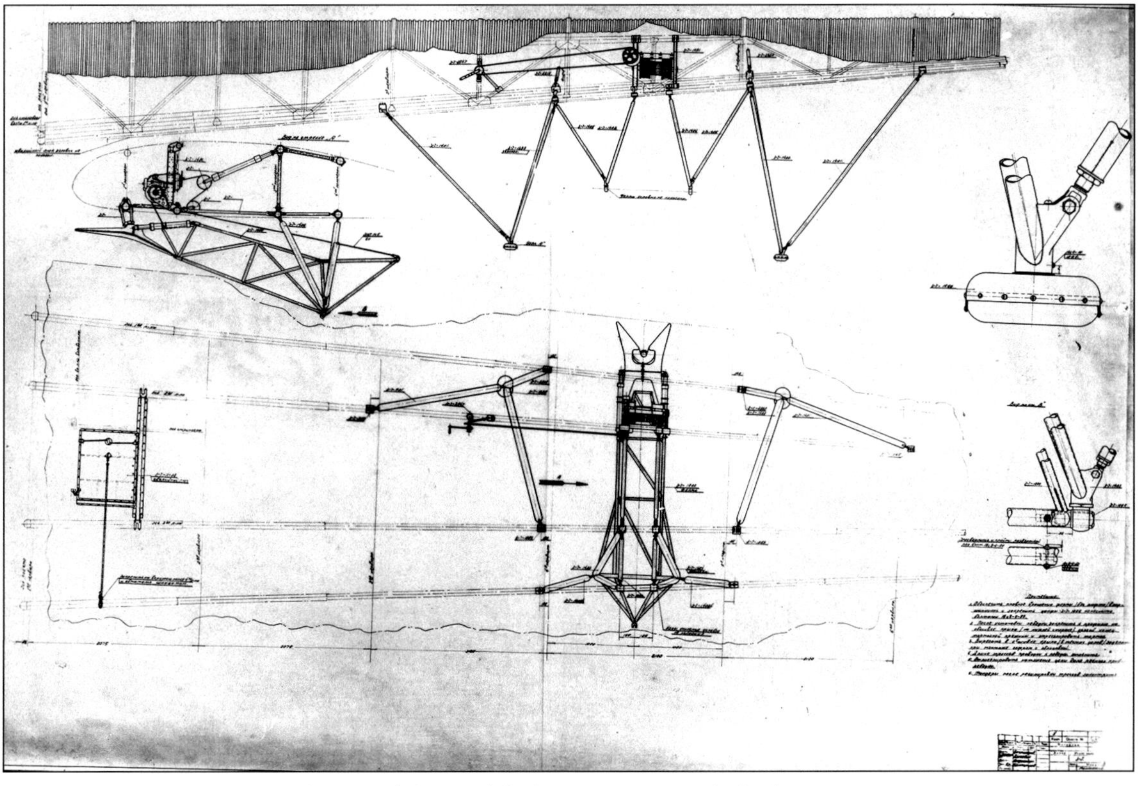

A general view of the modified Z-7 system, with the lowering truss.

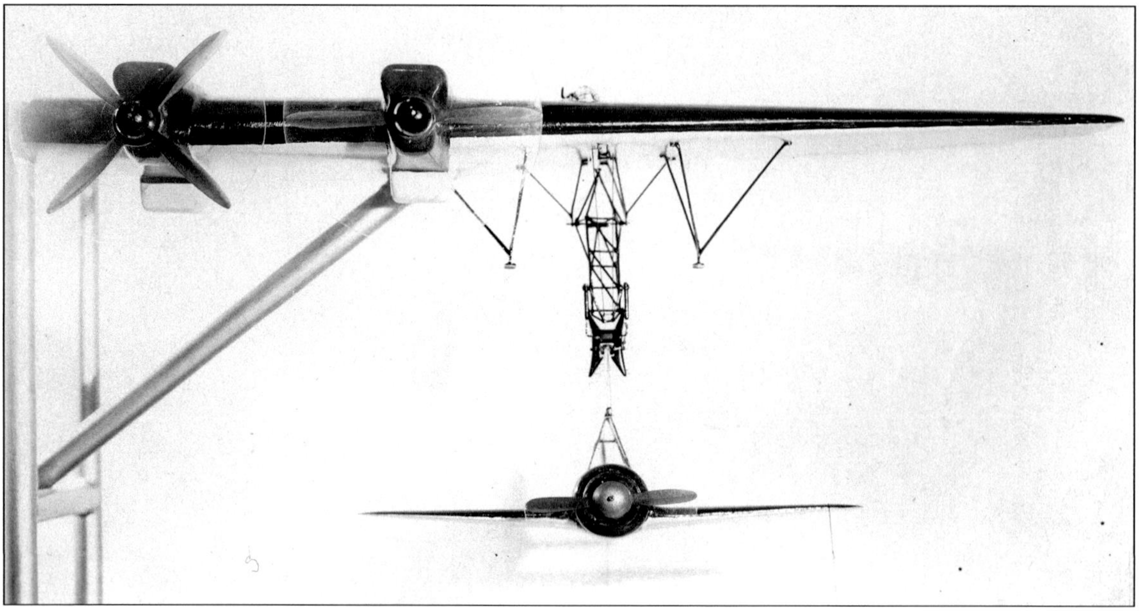

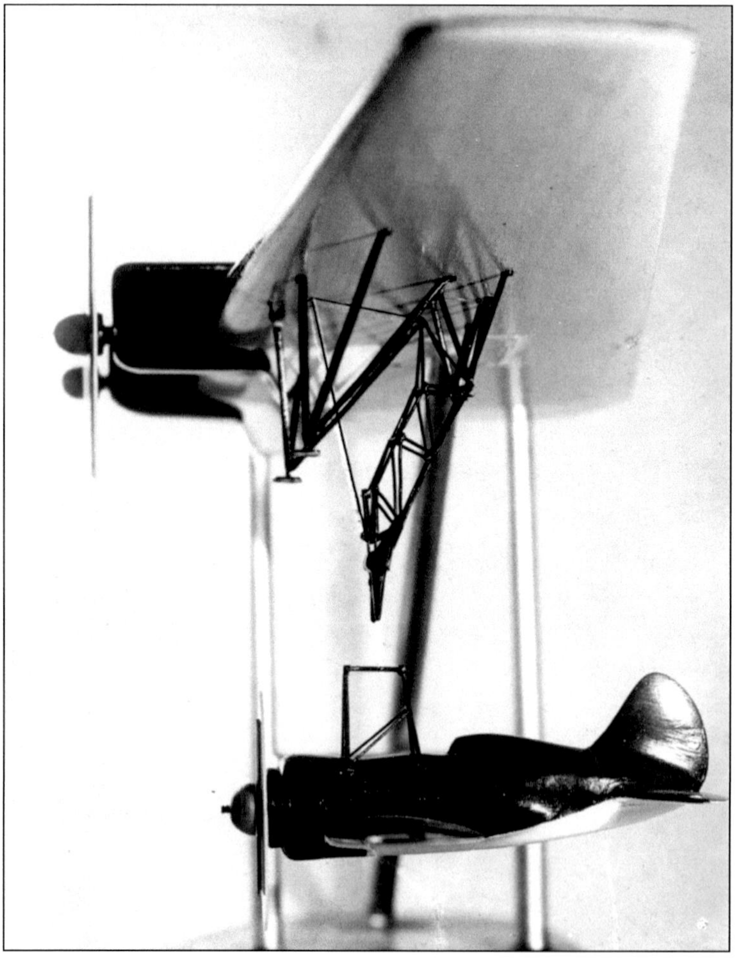

A demonstration model of the improved Z-7 configuration.

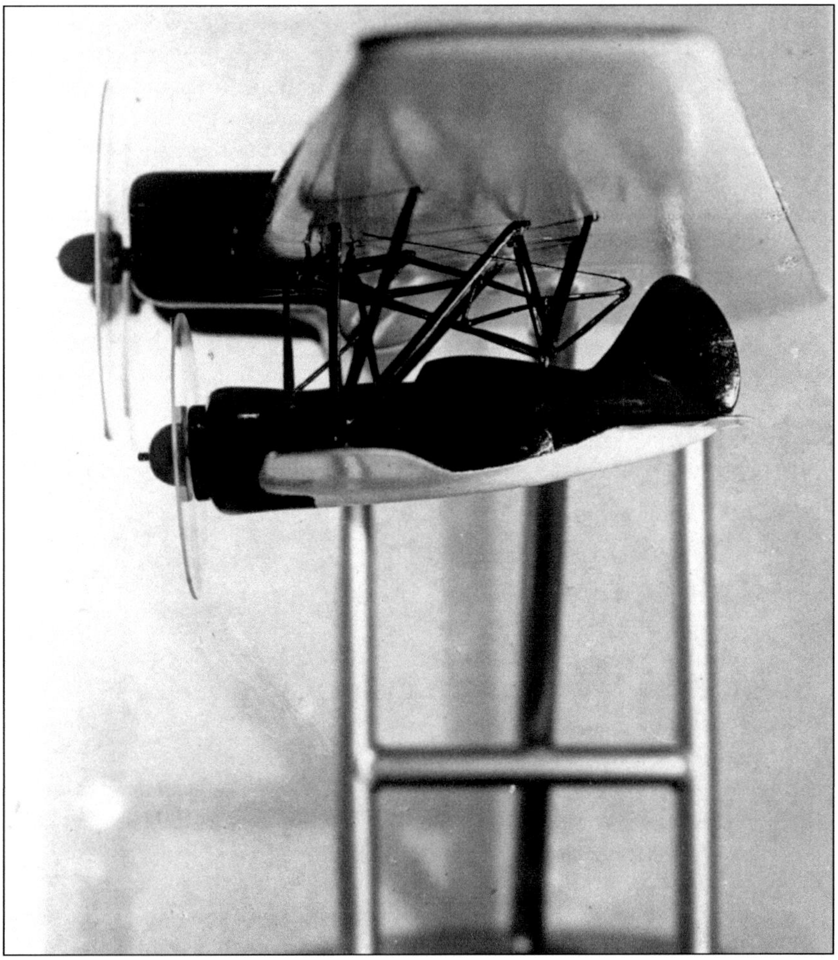

More photos of the demonstration model.

10

'Zveno-SPB'

The most famous combined-aircraft design by Vakhmistrov was referred to as 'Zveno-SPB' – and it consisted of a Tupolev TB-3 bomber and two Polikarpov I-16 fighters, which were each armed with a pair of 250kg bombs. Initially, the designer suggested using such suspended fighters as 'dive bombers' in 1936 after successful tests of the Z-6 system; it seemed that there would not be any extra hindrances for further specialization of the already well-tried attachment method. Indeed, the Zveno-SPB was constructed in the initial version and made its first flight in July 1937; however, the circumstances – the aviation industry standing and the situation in the air force – had considerably changed by the time in question.

The year 1937 in the Soviet Union is known for the beginning of mass purges among the leaders of the State and the army. In the middle of 1937, one of the key supervisors of the 'Zveno Aircraft' – Marshal Tukhachevsky – was arrested and shot; in the autumn, the Workers' and Peasants' Red Army Air Force Commander Alksnis shared the same tragic fate. At the time, many other responsible officials of the aviation industry and the air force became victims of repression. Among others, Vakhmistrov's elder brother, Boris, was arrested; later – and already imprisoned – he continued working together with A.N. Tupolev.

It is clear that these tragic circumstances considerably inhibited the decision-making in respect of the practical implementation of Vakhmistrov's inventions. Only a year later, in the summer of 1938, the *NII VVS* was able to conduct the evaluation tests of a system referred to as 'Tupolev TB-3 4M-34RN + two Polikarpov I-16 M-25s' ('Zveno-SPB'). The description of the parameters and capabilities of this configuration made a strong emphasis of the fact that it made efficient use of the obsolete TB-3 bombers and the widely-spread I-16 fighters acting as 'dive bombers'. Take-off, climb and flight were performed with all the six engines operative; in such a configuration, the surplus power as provided by the fighters was used to facilitate the TB-3's take-off. Each of the I-16s – armed with a pair of 250kg bombs – was intended to hit pinpoint targets, which were inaccessible for a high-altitude bomb attack. Once the bombs were dropped, the I-16s performed like conventional fighters – capable of protecting heavy bombers and returning to the base on their own. The I-16s' radius of action was increased by 50-80 percent, because when attached to the mothership, they were refueled from its tanks. For this purpose, the TB-3's outer wing tanks (600 liters each) were filled with B-3 gasoline, while the M-34 engines of the TB-3 bomber ran on B-2 aviation gasoline. For the bombing of close targets, the TB-3 did not even enter the enemy's air defense area; instead, it launched the I-16s above the 'friendly' territory. The I-16s with the suspended bombs had a maximum speed of 410km/h at an altitude of 2,500m. They could climb to 5,000m in 10.7 minutes and the service ceiling was 6,000m.

The explanatory note as to the tactical capabilities of the Zveno-SPB additionally stated that the configuration provided for the combined take-off and landing of the heavy bomber and the fighters attached under the wing of the TB-3. The fighters were secured to the TB-3's wing by means of two forward pyramids and the aft reclining brace using special locks. The same as in earlier Zveno configurations, the forward pyramids secured the I-16s via the locks on the fighter's fuselage. The aft brace was attached to fourth spar of the TB-3's outer wing; it was, essentially, a shaped steel tube with a fork to be secured to the aft ('back-mounted') lock on the I-16. Once the lock was released, the brace was reclined backward by means of a shock strut. When separating from the TB-3, the I-16's pilot disengaged the aft lock from the reclining brace and pushed the control stick. As a result, the I-16's angle of attack decreased and the forward locks released automatically once the aircraft turned through 3.5° – thus the fighter was launched into free flight.

The SPB configuration was provided with a two-way communication system, which included light signaling between the TB-3's captain and the I-16s' pilots and SPU-7 intercom – ensuring communication between the aircraft navigator, the I-16s' pilots and the rest of the crew. Besides, the TB-3 was fitted with a special periscope which turned through 360° and ensured the observation of the under-wing fighters. The light indication equipment included a signaling light (or a light annunciator) attached to the lower cap of the first spar on the TB-3, and a light annunciator near the I-16's pilot with 'Apply Power', 'Throttle Back' and 'Separate' placards corresponding to the instructions to increase/reduce engine RPM or to separate from the mothership, and 'Right', 'Left' and 'Up' placards informing the I-16's pilot of the enemy fighters' attack. Immediately before the command to detach, a horn sounded.

The Zveno-SPB prototype consisted of a Tupolev TB-3 bomber manufactured in 1937 with AM-34RN engines, and Polikarpov I-16 Type 5 fighters manufactured in 1938. For the I-16s, serial numbers 502 and 507 (i.e. s/ns 521502 and 521507) were specified. Following the State tests in 1938, the system was partially modified – in particular, the pyramids under the TB-3's wing were shortened from 1,295mm to 1,200mm, the design of bomb seats on the I-16s was changed and the fighters' refueling system was improved.

Marshal Voroshilov and People's Commissar for the Navy Frinovsky advocated accepting the Zveno-SPB into service as soon as possible. On 29 October 1938, they sent a letter to the chairman of the Defense Committee – V.M. Molotov – stating that '... the combination of the TB-3 with two I-16s could successfully be used in naval aviation to hit small-size targets, principally the enemy's fleet, and in land-based aviation to destroy bridges, electric power stations, etc'.*

The Draft Resolution of the Defense Committee of the USSR, which was enclosed in the above letter, reads as follows:

1. The Zveno-SPB to be accepted into service by the Workers' and Peasants' Red Army Air Force and the Workers' and Peasants' Red Army Naval Aviation;

2. The People's Commissariat for Defense Industry to provide, by 1 February 1939, the following number of aircraft for the manufacture of combined aircraft according to Comrade Vakhmistrov's method:
   20 TB-3 AM-34RN aircraft for the Workers' and Peasants' Red Army Air Force;
   20 TB-3 AM-34RN aircraft for the Workers' and Peasants' Red Army Naval Aviation;
   40 I-16 aircraft for the Workers' and Peasants' Red Army Air Force; and
   40 I-16 aircraft for the Workers' and Peasants' Red Army Naval Aviation.
   The Order to be placed at the facilities of Factories No.21 and No.32 of the People's Commissariat for Defense Industry;
   The development, manufacture, and installation of aiming devices for the dive bombers to be expedited;

3. The People's Commissariat for Defense to hand over, to the People's Commissariat for the Navy, 20 TB-3s to be converted into airborne motherships, within a month;

4. The People's Commissariat for Defense Industry to provide in 1939, for deployment of engineer Vakhmistrov's experimental work, Factory No.25, after freeing the latter from manufacture of consumer goods;

5. Comrade Vakhmistrov to provide, to the People's Commissariat for Defense and to the People's Commissariat for the Navy, a plan of further work.†

---

* Central State Archive of the Navy: collection r-1678; record statement 1, file 22, sheet 105.

† ibid.

The designer, inspired by this piece of success, indeed suggested several most interesting designs involving the TB-7 advanced bomber, MTB-2 and GST flying boats, the I-16 fighter with a 500kg bomb and the I-180 fighter.

Meanwhile, the year 1939 brought no perceptible changes in the sphere of practical operational use of Vakhmistrov's combined aircraft by the air force and the navy. The small design team was not moved from the premises of Design Bureau No.29 to other facilities, the TB-3s were not provided and Vakhmistrov – who in certain correspondence, was referred to as 'Z ('Zveno') Team Leader' – in practice held the appointment of an ordinary design engineer; however, at the time, the system of suspended aircraft continued to be modified – and in the summer of 1939, State tests of the configuration were conducted (the tests were performed using the same aircraft which had taken part in the 1938 flights). Captain Khripkov piloted the TB-3 as captain, with Fikson (Design Bureau No.29's pilot) as copilot. I-16 s/n 507 was piloted by Captain Nikolayev and I-16 s/n 502 was piloted by Senior Lieutenant Kubyshkin. In additional flights, the I-16s were piloted by the same officers; Major Lisitsyn piloted the TB-3 as captain, with 3rd Rank Military Engineer Fingerov as copilot.

First, in order to determine the reliability of all detachment mechanisms, thorough ground tests were performed; then, flights without the bomb load, but with further separation of the I-16s, were carried out. At the next stage, 130kg bomb cases were suspended under the I-16s and then dropped from them; finally, the intended FAB-250 bombs were used. It was noted that attaching four FAB-250 bombs to fighters took seven people and eight minutes, whereas the mounting of two I-16s under the TB-3 took 15 minutes with the participation of nine people.

According to the test report, the service ceiling of the Zveno-SPB was 6,300-7,000m; the maximum speed totaled 276km/h at an altitude of 3,000 meters and 271km/h at an altitude of 5,000 meters. The maximum speed of the I-16s with bombs suspended reached 379km/h at an altitude of 5,000 meters, and the service ceiling with the bombs was 6,800m. The speed at the moment of pull-down to dive was 165-220km/h. The pull-down was performed from an altitude of 3,500-4,500m, with diving angles of 30° through to 78°.

The gross weight of the Zveno-SPB configuration totaled 20,110kg without the bombs and 21,110kg with bombs suspended under the I-16s; the estimated gross weight of an I-16 Type 5 was 1,650kg. It was separately noted that for the period of the tests, the wing-located machine guns were removed, since they interfered with the mounting of bomb supports. The list of drawbacks included the hanging-down refueling hoses, which at the moment of the fighters' separation, were thrown back and broke the TB-3's skin.

For the building of the intended series of five Zveno-SPB combinations, it was suggested to eliminate the mentioned defects and to provide the prototype specimen for State tests. Furthermore, it was planned to use more advanced I-16 Type 24 fighters and to mount a large-caliber Berezin machine gun or a 20mm cannon in the middle turret. The report on the conducted joint operational verification tests – including all of the recorded objections and recommendations – was approved on 29 December 1939.*

To conclude this chapter, which is dedicated to the creation (and the improvement of) the Zveno SPB, it should also be added that in 1939, Vakhmistrov suggested using the I-16s with 500kg bombs. He believed that the new system – referred to as 'SPB BRAB-500' – would be a logical development of the 250kg design and that the bomb caliber should be increased to enable the system to hit fleet ships and concrete fortifications. An attachment fitting for the suspension of a 500kg armor-piercing bomb was mounted under the I-16 fuselage, and a special mounting with a PB-3 parallel-link motion mechanism was used in order to make the bomb avoid the zone of the rotating propeller; the PB-3 mechanism had already been tried out at the *TsAGI* and was even tested on the SB bomber. The new SPB BRAB-500 system was intended to be unified. It allowed using a FAB-500 bomb under the fuselage and two standard suspended 100-liter gasoline tanks

---

* A collection of documents relating to V.S. Vakhmistrov, file 5091/19: a report on verification tests of Zveno configuration 'TB-3 4M-34RN + two I-16 M25s' (Zveno-SPB) in cooperation with Design Bureau No.29.

under the wing of the I-16. Alternatively, two FAB-250 bombs could be suspended under the wing and two additional gasoline tanks under the fuselage.

Another modification comprised the mounting of a refueling hose on the aft reclining brace. The reason for the modification was that earlier, the refueling hoses had remained within the air stream and thus caused damage to the lower surface of the TB-3 wing. After the modification, the refueling hose – during the separation of the fighter – remained on the aft brace and disengaged automatically with the separation of the I-16.

According to calculations, the gross weight of an I-16 totaled 1,750kg with its wing machine guns removed and with the provisions for the installation of the PB-3; it reached 2,250kg with the bombs suspended and 2,460kg with the suspension tanks mounted. The full gross weight of an SPB BRAB-500 'Zveno Aircraft' totaled 23,710kg.

It is known that the SPB BRAB-500 was also designed for the TB-7 bomber in combination with the I-16, and such a configuration was included in the Plan of Experimental Work for the Aviation Industry for 1940-1941; however, such a configuration was never devised in detail, nor built.

A combination of a TB-3 4M-34RN + two I-16 M-25s – already referred to as 'Zveno-SPB' – at the *NII VVS'* airfield during joint tests in December 1939.

An I-16 rolled under the TB-3's wing. The winch-based lifting system has already been mounted; what remains to be done is to pull the cable, to close the locks and to retract the landing gear. As we can see, the fighter had to be slightly 'lowered' to attach the pyramid to the wing; for this purpose, small pits were dug under the wheels. Bomb cases are placed at a distance from the aircraft.

The aft lock on Frame 8 of the I-16 (in the open position).

The interface between the center wing section and the outer wing panels of the I-16, with the splice strip removed. The fairing of the machine guns, which were mounted in the wing before, is also removed; one can easily see the bomb attachment fittings.

An I-16 suspended on pyramids; FAB-250 bombs are mounted and ready for use.

A view of the interface between the outer wing and the center wing section of the I-16, with the splice strip and machine gun hatch removed.

The left forward lock for attachment of the I-16 on the pyramid mounted on the TB-3.

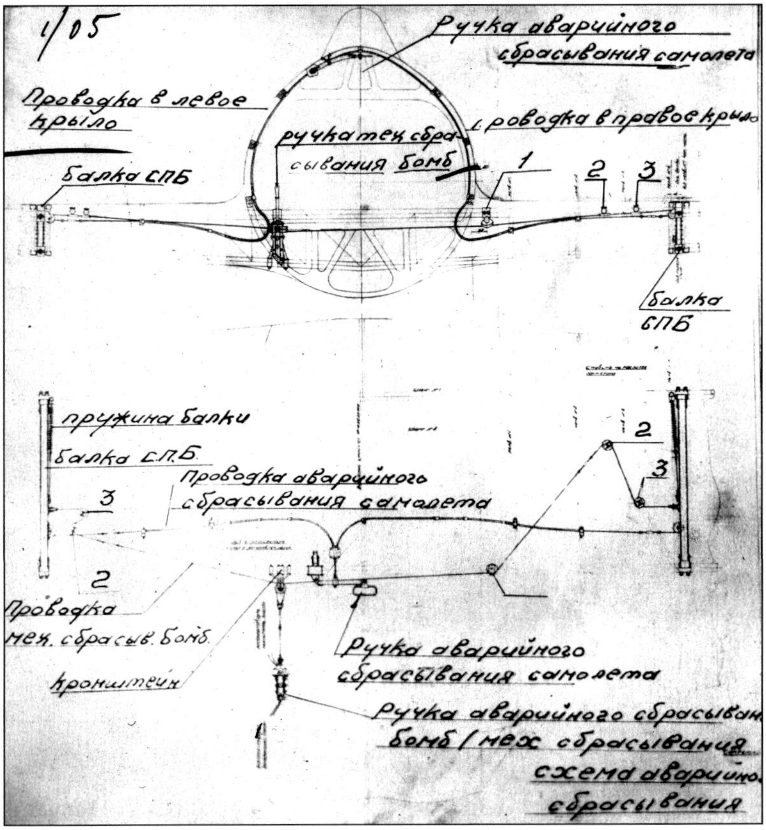

An original schematic for the I-16 launching and bomb drop control.

The central portion of the TB-3 4M-34RN fuselage (bottom view). The arrow shows the location of the periscope, which was used by the TB-3's crew to observe the suspended fighters.

A fragment of wing and the center wing section, near the location of the FAB-250.

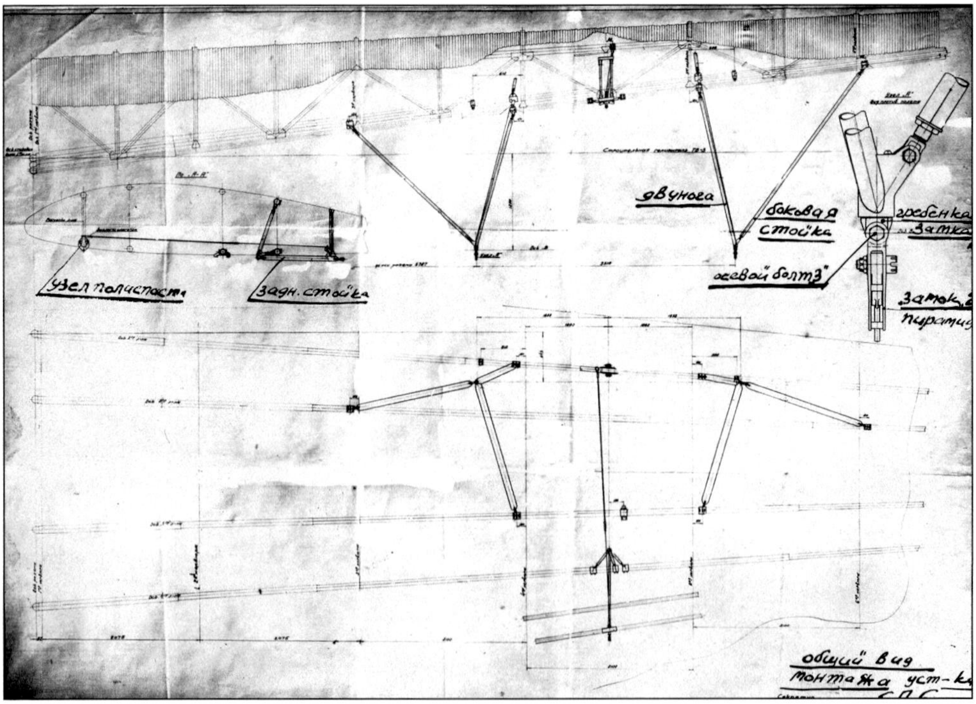

An original schematic for the installation of pyramids for fighters' suspension under the TB-3's wing.

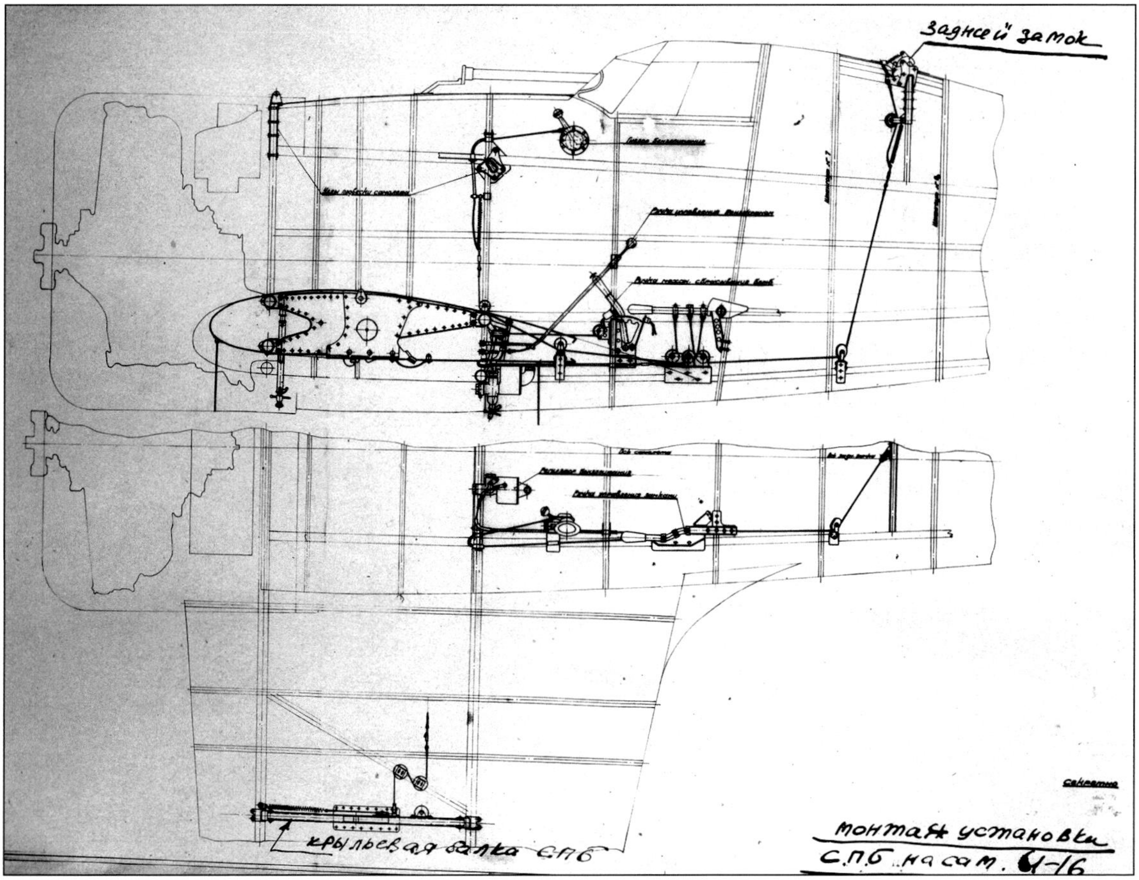

An original schematic for the installation of additional equipment on the I-16 SPB.

A FAB-250 bomb case under the I-16's wing.

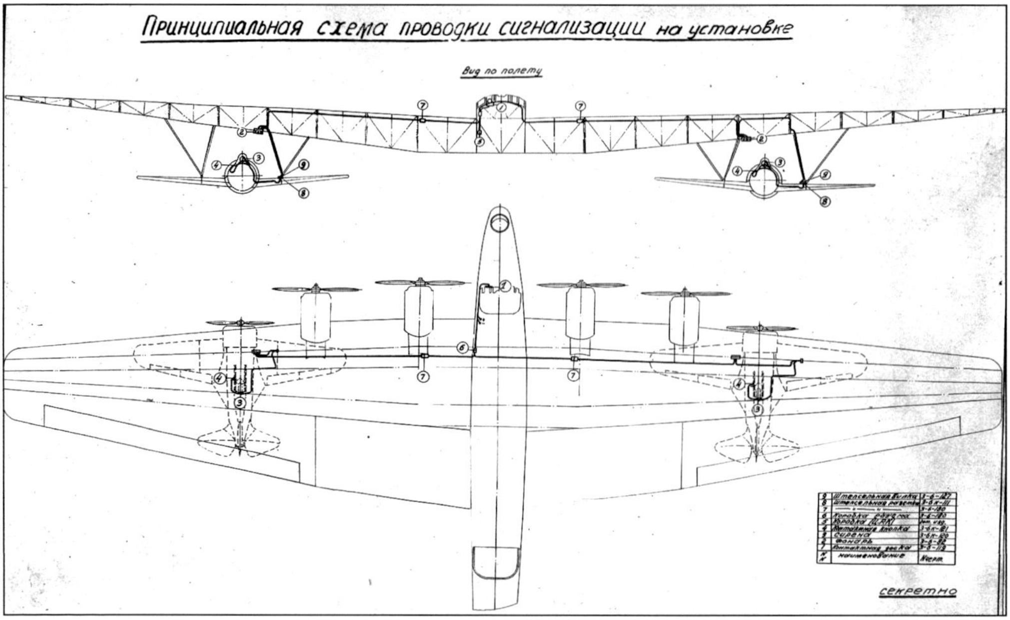

An original schematic for the signaling system wiring on the Zveno SPB combined aircraft.

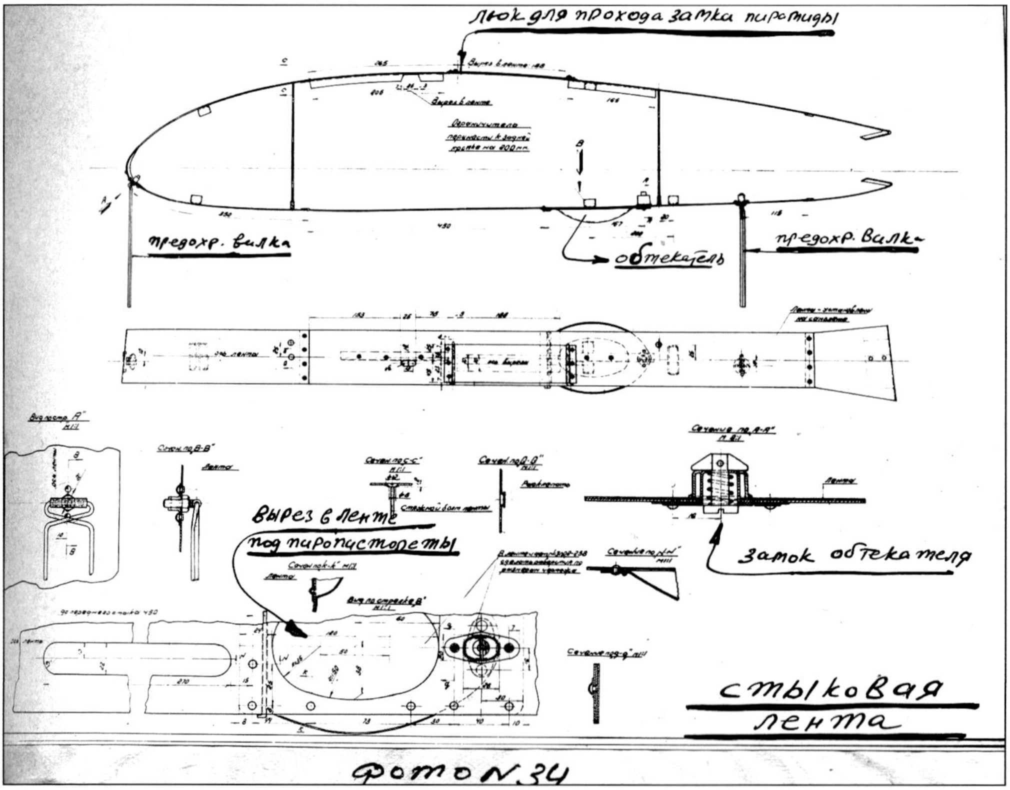

A strip covering the interface between the outer wing and the center wing section of the I-16, with cutouts for suspension fittings.

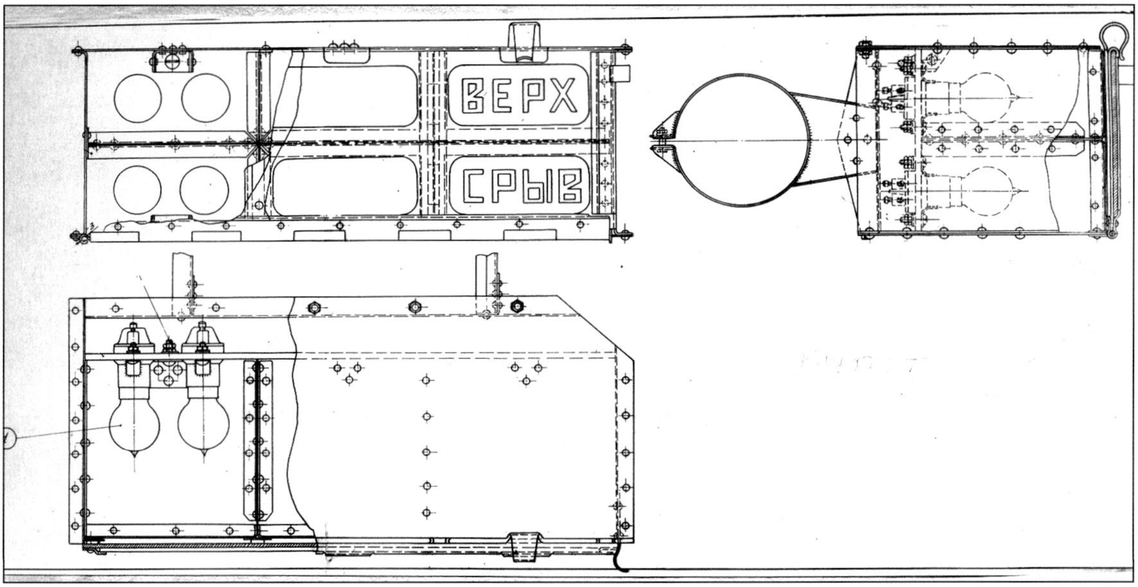

The indication light annunciator.

The periscope installed under the central portion of the TB-3's fuselage.

An I-16 being pulled up by a BL-3 winch for attachment to the suspension fittings.

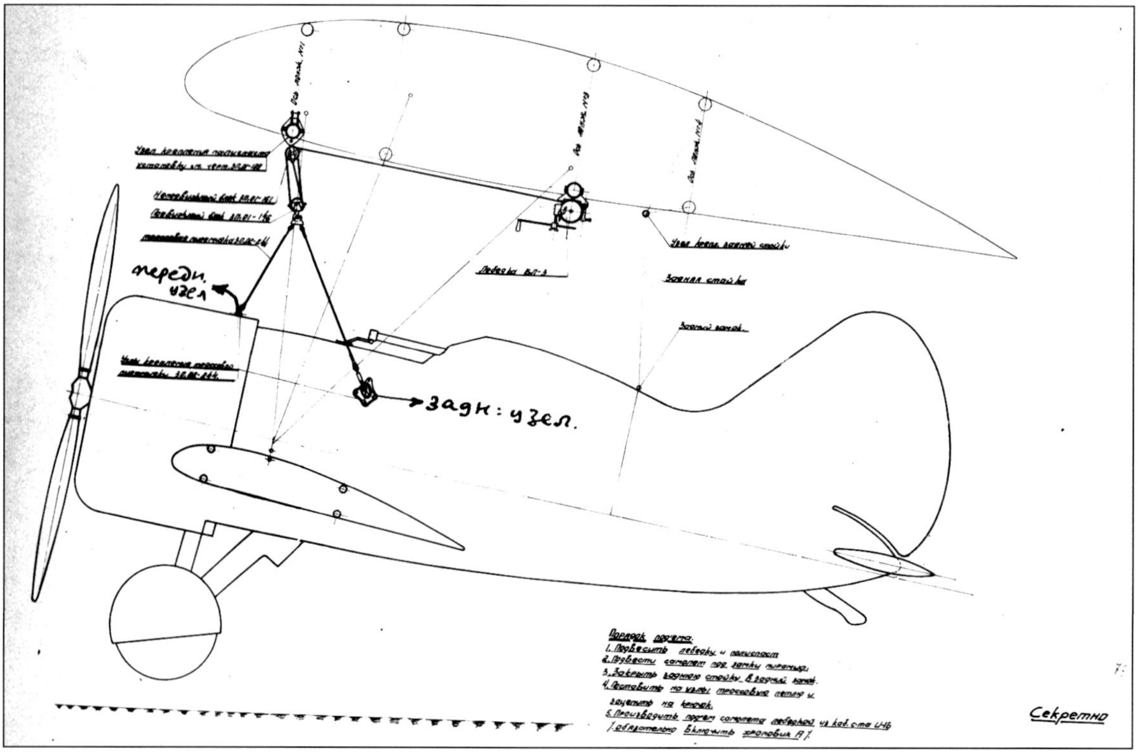

An original schematic for the attachment of an I-16 under the TB-3's wing.

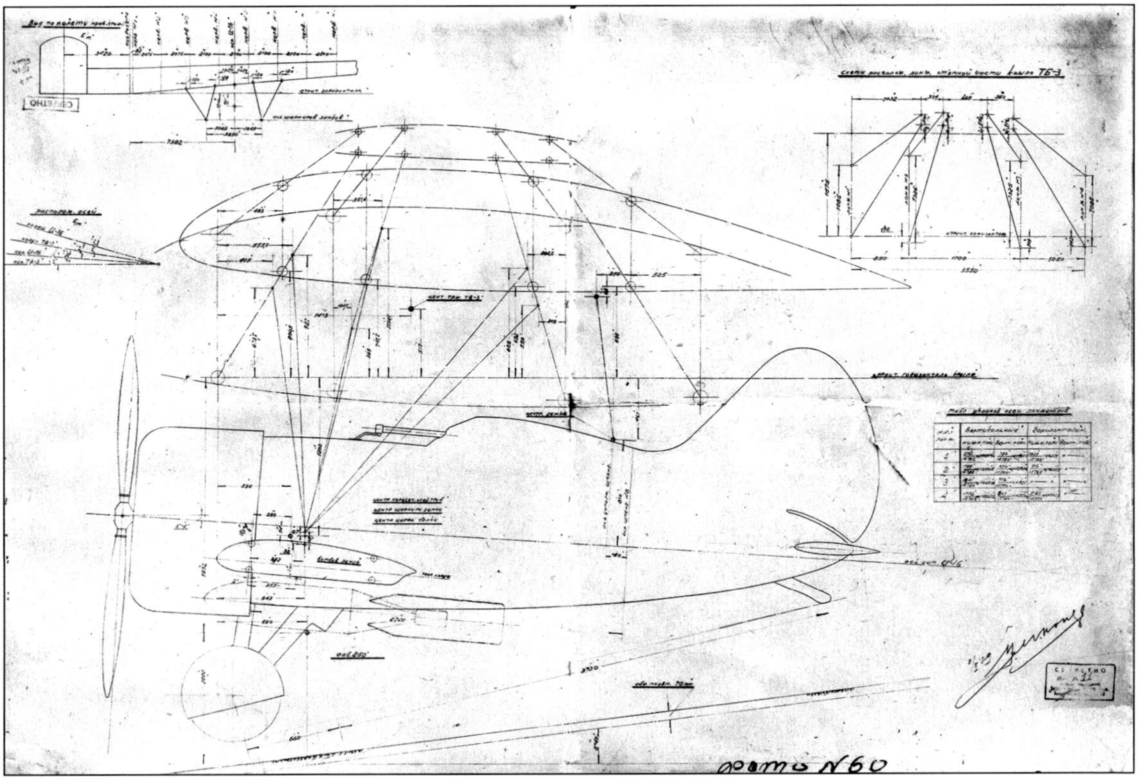

An original theoretical schematic for the attachment of the I-16 to the TB-3.

A combination of a TB-3 4M-34RN + two I-16 M-25s at the *NII VVS'* airfield during the tests in summer 1938.

An I-16 M-25 with two FAB-250 bombs under the TB-3's wing; one can clearly see the refueling hose.

A close-up view of the attachment fitting on the I-16's wing. Next to the attachment fitting is a connector for the intercom and the aural warning system connection.

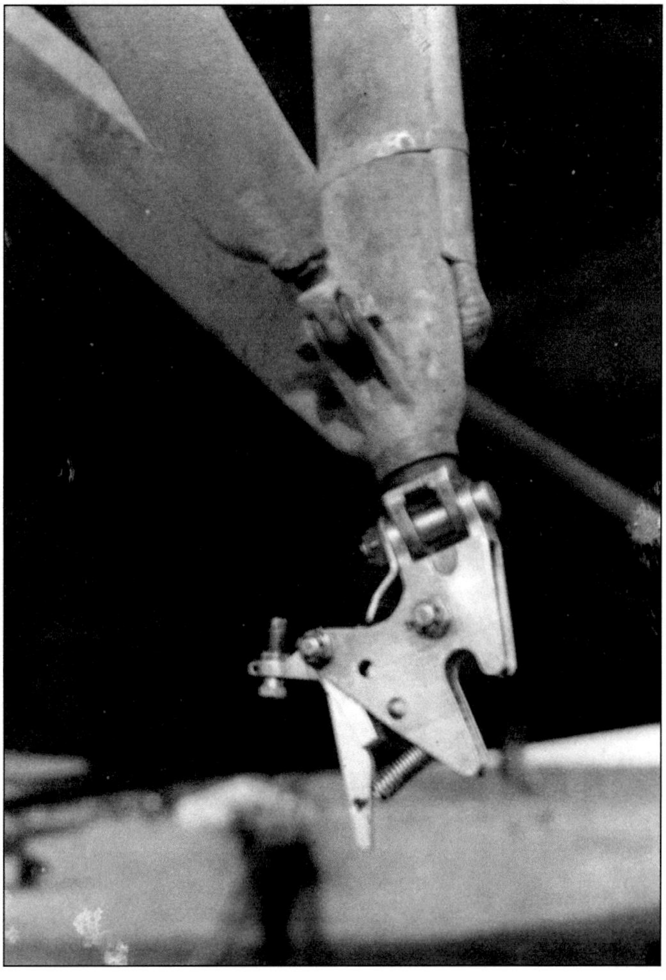

The attachment fitting on the truss in the open position.

The periscope under the TB-3's fuselage.

The FAB-250 bomb, with fuse locking secured.

The left I-16 in flight under the TB-3's wing; this photo was taken from the bomber's aft cabin.

The right I-16 in flight under the TB-3's wing.

An I-16 SPB with two FAB-250 bombs in flight.

The right I-16 separating from the TB-3.

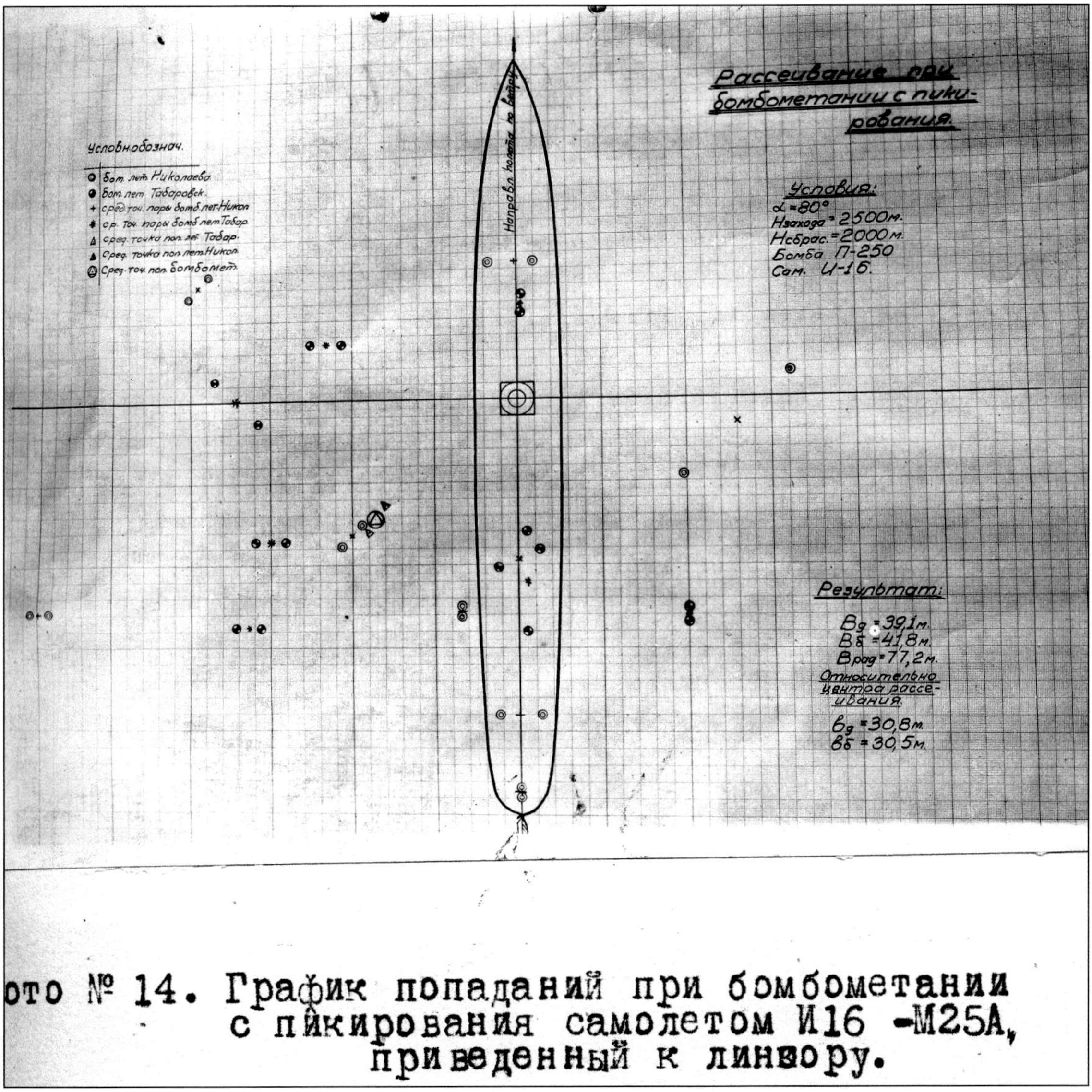

The results of hitting a battleship drawn on the ground; the hits made by Nyukhtikov and Taborovsky are shown (dated summer 1940).

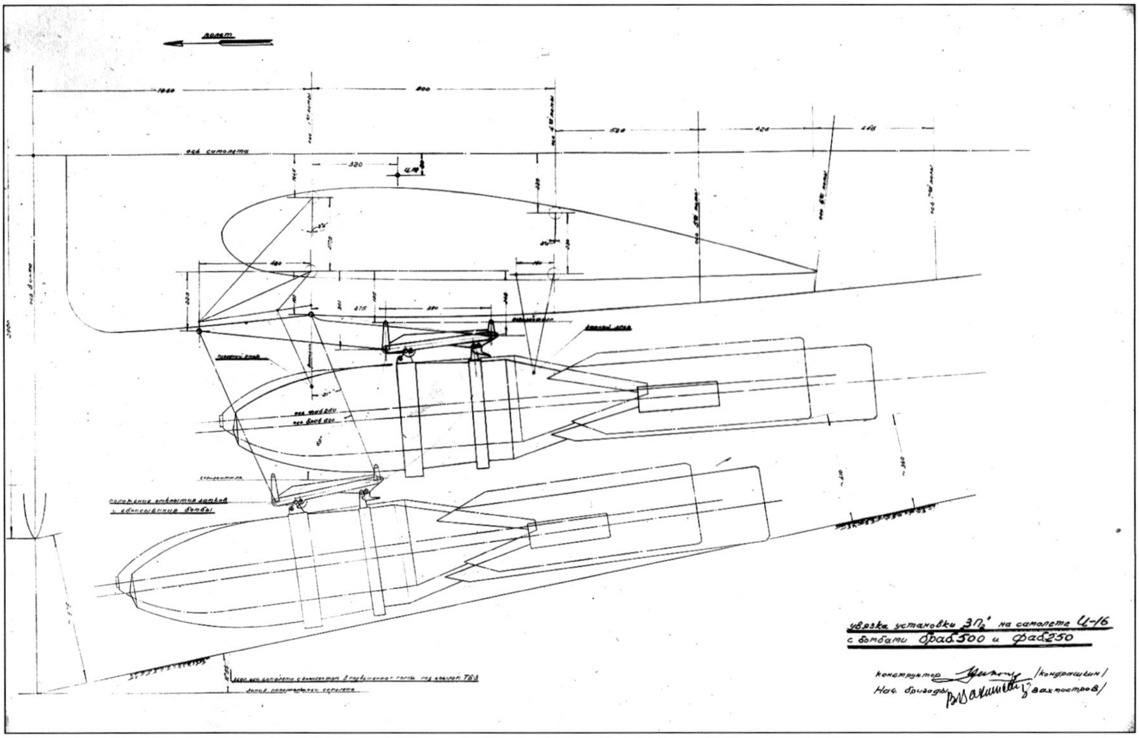

An original installation drawing for the suspension of BRAB-500 and FAB-250 bombs under the I-16.

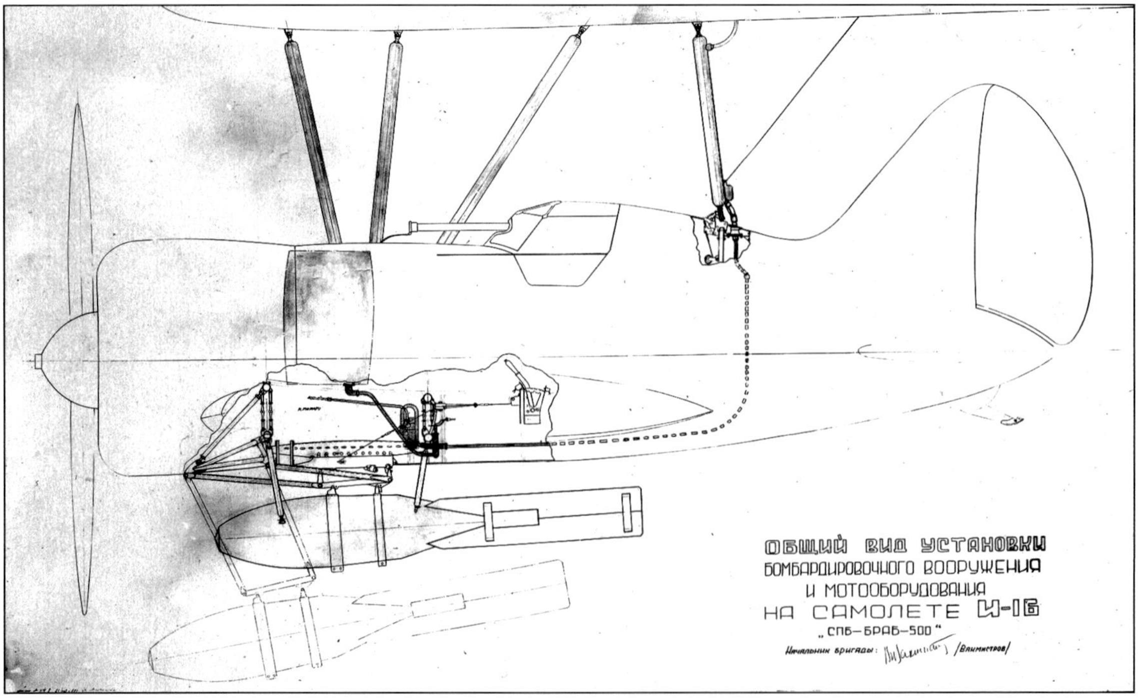

An original sketch of the SPB-BRAB-500.

11

Zveno-SPB: Mass Production

The period of tests of the Zveno-SPB system is reflected in the reminiscences of Vakhmistrov himself (these though are not quite clearly related to the dates):

> In 1938 or 1939, Voroshilov visited NII VVS to study the new aircraft. He was shown an 'avia-matka' lifting the I-16s with bombs. For a demonstration bomb-dropping, a battleship was made from canvas; the fighters destroyed it. Voroshilov liked everything very much. He talked to me and asked how the things were. I answered that things were going bad. Voroshilov promised to have a talk with People's Commissar of the Navy Kuznetsov. Then Voroshilov left the airfield, and I went to have dinner. Suddenly I saw a duty officer running to me. He said, 'Vakhmistrov is wanted on the phone'.
>
> It was People's Commissar of the Navy. He asked me to come and have a talk with him at midnight. That was when the work really got going.*

In fact, according to the available documents, the issue of ordering mass-produced SPBs arose in full scale in January 1939. During the first half of the year, the issue was not settled in practice; at the time, Vakhmistrov's small team was only working out the drawings for mass production. Meanwhile, it looked as if things had got going – and on 22 February 1939, the Directorate of Naval Aviation addressed a request to the Directorate of the Air Force of the Workers' and Peasants' Red Army to provide twenty TB-3 4AM-34RNs to build SPBs. Simultaneously, a request was sent to the Technical Directorate of the People's Commissariat for Aviation Industry to provide production facilities for the placing of the order to build 20 SPBs. Actually, it turned out to be rather difficult to obtain such a number of aircraft, though referred to as 'obsolete'; for half a year, nothing happened. Only on 20 October 1939, the air force management ordered to hand over twelve TB-3RNs from the Baltic Fleet, the Black Sea Fleet and the Pacific Fleet (four aircraft from each fleet); however, this order was not entirely fulfilled. The documents confirm only the initial handover of three TB-3 4AM-34RNs.†

To fulfill the order, the People's Commissariat for Aviation Industry assigned Factory No.207 in the settlement of Dolgoprudny, in the Moscow area. On 3 August 1939, the factory sent to the Navy's Aviation Directorate a draft contract for the manufacture of five SPBs. It is known that the terms for delivery of the equipment for the manufacture of SPBs (as suggested by the factory) were, in the customer's opinion, hardly acceptable; in particular, even the initial working schedule implied that the work would be complete in the first quarter 1940. The reason for such an unfavorable attitude to the SPB was that Aviation Factory No.207 (formerly *Dirizhablestroi* - Airships Manufacturer), which was rather a kind of experimental facility, had earlier received an order for the mass manufacturing of the BB-1 (Su-1) short-range bombers – therefore, one

* Archive of M.E. Zhukovsky Memorial Museum: a collection of documents relating to V.S. Vakhmistrov; a record of V.S. Vakhmistrov's interview with M.S. Arlazorov on 7 February 1969.

† Central Archive of the Navy: collection r-61; record statement 3, file 2: 'On Implementation of SPB Aircraft Designed by Engineer Vakhmistrov in Naval Aviation'.

can well understand the striving of the director of Factory No.207 (Gorin) to make every effort to refuse to receive and fulfill the new task.

Meanwhile, the topic of the 'Zveno Aircraft' was given a renewed impetus in early 1940, in connection with the events at Karelian Isthmus. The Soviet-Finnish War, in which the Red Army (at first) experienced certain ill success, made the leaders of the State think again as regards some of the military inventions. In view of the situation, Design Engineer Vakhmistrov (obviously, he had not been appointed a chief designer, nor even a team leader) sent a letter to the People's Commissar of Defense, Marshal Voroshilov.

The letter read:

> The most efficient way to destroy small-size targets in the enemy's deep rear (e.g. ships, bridges, warehouses, etc.) is the dive bombing. One of the dive bomber types that we have at our disposal is the Polikarpov I-16 fighter fitted with a system, which I designed. The system allows an I-16 suspended under the TB-3 bomber to carry two FAB-250 bombs. The TB-3 can bring two I-16s with bombs to the desired location at an altitude of 5,000 or 6,000 meters with a speed of 220-250km/h, launch the I-16s, and return to the base. The TB-3's radius of action under such circumstances can be 600 to 800km.
>
> The I-16s are refueled during the combined flight from the TB-3's tanks, and are launched with a full fuel reserve. The tactical range of the I-16 will thus be approximately 400km.
>
> An I-16 M-25A with bombs has a speed of 410km/h at an altitude of 2,500km, and service ceiling of 6,800km. Flight performance of the I-16 M-63 should be, accordingly, even better.
>
> Once the bombs are dropped, an I-16 behaves like a conventional fighter, and is able to defend itself.
>
> Currently, Factory No.207 is finishing the manufacture of five Zveno-SPB equipment kits; this equipment can be promptly installed to combine five TB-3s and ten I-16s into Zveno-SPB configurations. If appropriate orders are given to the industry (Factory No.207 and Design Bureau No.29), the combined aircraft can be ready in seven to ten days from the date when the TB-3s and the I-16s are received. It will be better if assembly work is carried out at Factory No.23 in Leningrad by the teams of Factory No.207 and Design Bureau No.29.*

Indeed, equipment for the Zveno-SPB system was being manufactured at full speed – and the assembly team was being formed in order to be delegated to certain facilities in the Leningrad area to install such equipment; however, as is often the case, it turned out to be too late: in March, the military activities terminated and the debut of the I-16 dive bombers did not take place in spring 1940.

Meanwhile, the new Aviation Factory No.289 of the People's Commissariat for Aviation Industry was arranged on 5 March 1940 at the facilities of Design Bureau No.29 in Podlipki. Vakhmistrov – after a considerable period of time – was appointed head of the team, which (under designation 'Z') was included among the design teams of the factory. The factory, at the time, continued to work on high-altitude BOK aircraft designed by specialists of the Special Design Bureau, or *Byuro Osobykh Konstruktsiy*. Also, the factory became a principal place of work for designer Sukhoi and Privalov's team (the latter was dealing with troop-carrying equipment).

Somewhat later, the Scientific and Research Institute of the Red Army's Principal Directorate for Aviation Procurement (*NII GUAS KA*) started training crews to operate and carry out maintenance of the Zveno-SPB system. Two flight and maintenance crews (one crew from the Black Sea Fleet and one crew from the Baltic Fleet of the Red Banner) arrived at the facilities of the Research Institute – and the training sessions were reported to have started on 5 June 1940; a note to readers: all further mentions of the operation of suspended dive bombers are connected solely with the Black Sea Fleet.

---

* ibid.

This was just the right moment for the said training flights – taking into account the fact that the tests of the prototype mass-manufactured Zveno-SPB configuration (comprising TB-3 s/n 22731 and I-16s s/n 2421и11 and s/n 2421з68) were completed on 23 June 1940. The acceptance tests were performed from 17 to 21 June 1940 at *NII VVS'* airfield in Chkalovskaya, in accordance with a program agreed upon with Factory No.207 and the Navy's Aviation Directorate. Colonel Fomichev was the navy's flight group commander and the copilot of the TB-3; Captain Razumov piloted the TB-3 (as the left pilot). The I-16s were piloted by Lieutenant Litvinchuk and Lieutenant Ryzhov. Throughout the tests, six attachments of the I-16s to the carrier aircraft were performed. Five mid-air separations were performed upon command from the TB-3, and one separation was carried out by means of an emergency launcher (on the decision of the I-16s' pilots). In total, three flights with bombs carrying and dropping were performed – and all the separations were performed at a speed of 200km/h. One of the flights was carried out with the FAB-250, with separation at an altitude of 3,000 meters, followed by a dive at an angle of 85°. All the locks demonstrated failure-free operation in all cases of docking and separation. It was noted that the I-16s were fitted with signal lights mounted in front of the aiming devices; such signal lights indicated when the pilots were being called by the TB-3 captain. Also, it was suggested that these signal lights should be further installed on all I-16 SPBs instead of the horns, with which such aircraft had been previously fitted.

In the opinion regarding the results of the tests, a need to perform certain minor modifications (such as reinforcing the springs on pyramid locks) was pointed out; there was also a suggestion that a special trolley should be arranged and used for the lifting of bombs to the I-16s.

Certain peculiarities of the fighters' in-flight behavior were also noted; for example, the 'settling-in' altitude at separation from the TB-3 at a speed of 2,000km/h with 250kg bombs was 500 meters. Speed at pull-down was 250km/h; speed at pull-out – at an angle of 85° – was 480km/h. The test report also stated that the system had passed all tests successfully, and was found fit for operation.

Upon finalizing all the required formalities, on 21 August, Zveno-SPB was to have departed (in full configuration) to its intended base within the Black Sea Fleet; however, an unexpected thing happened. Lieutenant Junior Ryzhov, who had, the day before, been on leave in Moscow, got involved in a romantic affair and showed up late. Of course, the young pilot was reprimanded for having upset the plans for the departure; however, now we know that the first Zveno-SPB was received by the Black Sea Fleet in late August 1940. No documents have been found witnessing the time of arrival of the other two complete Zveno-SPB combinations, so we may only suppose that they were assembled during the autumn.*

Before the end of the year 1940, the navy's management remained interested in Vakhmistrov's systems. This is confirmed by the document dated 11 December 1940 and entitled 'Operational and Tactical Substantiation of Using the Combined 'TB-7 + two I-16s' *Zveno* Aircraft in Naval Aviation'.

The following was specified among the principal objectives:

1. Covering the actions of the friendly fleet from the enemy's air attacks within a range of up to 300km from the nearest airfield;
2. Covering the Naval Bases from the enemy's aircraft by means of long-time air patrolling (for 6.5 hours).
3. Dive bombing of the enemy's transport vehicles and light forces within a range of up to 300km from the nearest land-based aerodromes;
4. Protection of our communications (the mouth of the White Sea, Murmansk, Vladivostok, Sovetskaya Gavan) against the enemy's air strikes.

The key advantages of using the combined 'Zveno' aircraft, in comparison with the fighters in current service, which act directly from the aerodromes, are as follows:

* Central Archive of the Navy: collection r-61; record statement 4, file 363, ⊠.199.

1. The use of the TB-7 as a mothership for the fighters increased the existing fighters' range twice.
2. The mothership enables the fighter aircraft maintain, with less stress, continuous air patrolling throughout the daytime for the purpose of guarding naval bases or other critical facilities (Example: For naval base defense by means of simultaneous air patrolling as carried out by 18 fighters operating from the aerodrome throughout the 12 hours of daytime, 144 sorties 1.5 hour long each would be required). The use of a 'Zveno' aircraft enables the fighter aviation to ensure, with less stress, escorting the fleet ships within a range of up to 300km from the nearest aerodrome.[*]

Meanwhile, the fate of further improvements to Vakhmistrov's systems were determined as early as summer 1940. At the time, the aviation industry experienced an abrupt reduction of experimental and research work. The oncoming war was clearly felt – therefore, the management strived to make the aircraft designers focus their efforts on more 'realistic' activities. The funding of 'high-altitude' projects (stratospheric aircraft, pressurized cabins and turbo-compressors) at Aviation Factory No.289 was cut off. As for the suspended aircraft, on 19 June 1940, People's Commissar of the Navy Admiral Kuznetsov sent a letter to Marshal Voroshilov, which read as follows: 'Under my Order No.00154 and No.00155, all experimental work relating to glide torpedoes (at Factory No.379) and to SPB utilizing the platform of TB-3 4AM-34RN (Comrade Vakhmistrov's design) have been stopped'[†] – thus the story seemingly ended at the most interesting point; however, Vakhmistrov kept supervising his topic until the end of 1940 – at least in terms of technical support for the already operational SPBs. Additionally, there is an impression that up until mid-1941, his activities were not quite clearly specified. Under the circumstances, there took place (rather unexpectedly) a junction of creative aspirations of Vakhmistrov and Polikarpov in the area of creating a special combat troop-carrying glider – known under the designation 'BDP'.

The date 23 January 1940 should be deemed a starting point in the history of Soviet multi-seated unpowered combat aircraft; on that very date, the Directorate for Manufacture of Troop-Carrying Gliders was created within the People's Commissariat for Aviation Industry. In autumn that year, at a special meeting (which was attended – among others – by I.V. Stalin), a decision was made to hold a 'tender' to select the best glider specimens and then to launch mass production. For the 'tender', gliders were created by O.K. Antonov, V.K. Gribovsky, P.V. Tsybin and N.K. Kurbala. The most successful ones (the A-7 and the G-11) entered mass production and were successfully used in World War II.

Nikolai Nikolayevich Polikarpov had never before taken part in the design of unpowered flying vehicles; his suggestion of a troop-carrying glider was presented in early June 1941 and, by all appearances, came as a surprise for a lot of specialists. For many years afterward, aviation industry officials and experts kept wondering why the renowned designer had decided to 'play on other people's ground' and deal with an *a priori* secondary-importance design – trying to understand the topic that was somewhat on the sidelines of mainstream aviation development… Indeed, the gliders were intended to be manufactured not even at aviation factories, but at peripheral production facilities such as woodworking and furniture factories. Seemingly, the reason for such a strange interest lied in the fact that Polikarpov had his own vision of the problem and was contemplating something new and original, but the real reason was different: it was the extravagant and indefatigable Vakhmistrov who recollected his quite recent interest in unpowered flying vehicles. It was he who suggested creating a glider not just for cargo or troop transportation in the rear (regardless of whether it be friendly or the enemy's rear), but rather as an offensive means. Matching the dates of the respective events, we can suppose that the direction of the designer's thinking was influenced by what happened in late May 1941. At the time, fascist Germany attempted a seizure of the island of Crete, which was guarded by the British troops. The attack was made from the air, with troop-carrying gliders also participating. The goal was achieved in spite of the quite tangible casualties suffered by the Germans. Analyzing the operation in terms

* Central Archive of the Navy: collection r-61; record statement 7, file 102, л.55.

† Central Archive of the Navy: collection r-61; record statement 7, file 47, л.175-176.

of participation of aircraft, Vakhmistrov came to the conclusion that the attacking party could avoid high casualties if it used special-purpose, well-armed and armored landing aids. In accordance with this concept, he suggested to Polikarpov that they should design a special-purpose glider. The troopers staying inside it could deliver fire against the ground-based enemy force by using their personal weapons. Sufficient protection in the air was provided by on-board removable armor plates, which were then (after landing in the open country) used as shields for the sheltering of the troopers.

The glider, which was designated 'Combat Troop-Carrying Glider' *(Boyevoy Desantny Planer*, or *BDP*), was carried out by Polikarpov's design office (with the close participation of Vakhmistrov) throughout June 1941. The design project of the glider, which was internally referred to within the design office as 'Item C', was completed several days before the fatal date of 22 June. It was, essentially, a high-wing monoplane with a 20m wing span and a smooth spindle-shaped fuselage. The glider had an all-wood structure, so that it would be as cheap and easy to manufacture as possible. The size of the glider, which was capable of carrying 16 armed soldiers, was selected on the assumption that it would be towed by the SB bomber and the DB-3 bomber. It was also expected that the soldiers' small arms would be usable to repel the enemy fighters' assaults on the 'flying train'.

The *BDP* design was approved by the aviation industry management and accepted for implementation at Aviation Factory No.51. By the late summer of 1941, the glider was built and prepared for flight tests. Flights of the first prototype *BDP* started in September and were performed at the airfield of the Flight Research Institute (*Lyotno-Ispytatelny Institut*, or *LII*) in the Moscow suburbs. We will not attempt to trace the *BDP's* fate in this study, since it has already been described more than once; we will only note that one of the principal initiators of this extravagant flying vehicle returned to his 'suspended dive bombers' as early as July 1941.

12

Zveno-SPB: Participation in Combat

Several documents state that in total, five equipment sets were manufactured and used for the Zveno-SPB system; however, three TB-3s with AM-34RN engines were initially allocated for re-equipment. It was exactly those three TB-3s which were accepted into service by the 2nd Special Squadron of the 32nd Fighter Regiment of the 62nd Aviation Brigade of the Black Sea Fleet Air Force stationed in Eupatoria. The 2nd Special Squadron was under the command of Arseny Vasilyevich Shubikov – an experienced pilot, as well as being a participant of the war in Spain and the holder of several orders.

Until the end of the year, the 2nd Special Squadron was actively practicing the tactics of operating the fleet's new arms materiel, with military ships of the potential enemy being the principal targets. In January 1941, however, the command of the fleet adopted a less favorable attitude to the extravagant configurations, which were duly dubbed 'Shubikov's Circus' (after the name of the squadron commander), and ordered that the combinations should be dismantled. They were again made use of only in the summer of 1941, with the war already in full swing.

In the beginning of the war, the Black Sea Fleet aircraft were repeatedly conducted bombing strikes on the territory of Romania, which (simultaneously with Germany) had attacked the Soviet Union. The railroad bridge over the Danube near Cernavodă station – 60km to the west from Constanţa – was considered one of the most important targets for the bomber aircraft. This was, essentially, major bridgework, with its total length of 1,662m and a total height of 75m at the highest point; the bridge carried, under its lower deck, the Ploieşti-Constanţa oil pipeline. Destruction of the bridge would entail the stoppage of all rail traffic between Bucharest and other industrial centers, the break-up of principal transport communications with Constanţa port and the weakening of supply to Romanian troops. Repeated attempts by the 63rd Bombardment Brigade of the Black Sea Fleet to destroy the bridge with the SB and DB-3 bombers were unsuccessful. Under the circumstances, it occurred to the command of the fleet that there were special-purpose Zveno-SPB systems among the fleet's materiel – and consequently, a decision was made to restore Vakhmistrov's systems as soon as possible for further operational combat use.

On 22 July 1941, Black Sea Fleet Air Force Commander Major General Rusakov received a message that read:

> People's Commissar of the Navy Admiral Kuznetsov has authorized the use of engineer Vakhmistrov's Zveno-SPB combinations. Comrade Vakhmistrov is delegated to provide technical guidance for the mounting of the appropriate equipment on the TB-3 and the I-16 aircraft. Besides, Comrade Vakhmistrov will be in charge of initial operation of Zveno-SPB until the flight crews and maintenance personnel learn how to operate and maintain the Zveno aircraft. It is hereby ordered to report the end of assembly activities and the results of in-flight check of attachment/detachment mechanisms. Signed by: Deputy Head of the Naval Air Force Major General for Aviation Korobkov.*

* Central Archive of the Navy: collection r-61; record statement 7, file 47, л.175-176.

According to the fact that the above message states that 'People's Commissar... has authorized...', it should be assumed that the preparation of the 'Zveno Aircraft' for military operation by the time had already started; furthermore, they were almost ready for operational use.

We know the principal participants of operations involving Zveno-SPBs during the hard summer of 1941:

- Heavy TB-3 carriers: Flight Commander Major E. Rozenko
- TB-3 : Captain Senior Lieutenant Serafim Petrovich Gavrilov
- TB-3 : Captain Senior Lieutenant Nilkolai Ognev
- TB-3 : Pilot Lieutenant Trushin
- Captain Arseny Vasilyevich Shubikov: Commander of the 2nd Special Squadron and an I-16 pilot
- Lieutenant Evgraf Mikhailovich Ryzhov: An I-16 pilot, Lieutenant E.M. Ryzhov did not take part in operational use of the SPBs. Before the war, he had mastered the MiG-3 and flew this fighter during military operations of the 32nd Fighter Regiment. On 23 October 1942, he was awarded the title of the Hero of the Soviet Union for courage and valor. In total, he made 300 sorties and shot down 11 enemy aircraft personally and six enemy aircraft in a group with his comrades
- Lieutenant Boris Mikhailovich Litvinchuk: An I-16 pilot, Lieutenant B.M. Litvinchuk was fighting throughout the entire war; he made 450 sorties, won 18 air battles and was awarded the title of the Hero of the Soviet Union
- Senior Lieutenant Boris Ivanovich Filimonov: An I-16 pilot
- Lieutenant Alexander Vasilyevich Samartsev: An I-16 pilot
- Lieutenant I.Y. Kasparov: An I-16 pilot
- Lieutenant Pavel Grigoryevich Danilin: An I-16 pilot
- Lieutenant Dmitry Ivanovich Skripnik: An I-16 pilot
- Sergeant S. Kuzmenko: An I-16 pilot

On the same day (22 July 1941), an order from Moscow followed to commence the operational use of 'Zveno Aircraft'; the first sortie took place on 26 July and an attack was made on Constanţa port structures.* At daytime, a couple of TB-3s approached the Romanian coast – and at a distance of 40km, launched four I-16s. The fighter element – comprising Captain Shubikov's and Lieutenant Litvinchuk's I-16s – approached the city and, in the presence of the Messerschmitt aircraft that were loitering nearby, made two leisurely circles around the target (an oil refinery) and then dropped the bombs. Another fighter element (the I-16 fighters piloted by Filimonov and Samartsev) bombed a floating dock; then, all four I-16s uneventfully left the attack area and flew away at an increased speed. They made an intermediate landing in Odessa, refueled and flew to Eupatoria. It should be noted, however, that one of the I-16s failed to reach Odessa because of a lack of fuel – and therefore had to make an emergency landing. In view of the incident, a decision was made to promptly equip the SPB's I-16s with an additional under-fuselage fuel tank of up to 95 liters for further sorties.

For the Cernavodă bridge raid, as many as three Zveno-SPBs were prepared – and the mission took place on 10 August at 0300 at night. All three prepared TB-3s departed; however, one of the carriers flew half the required distance and then had to return to base due to mechanical problems. At 0510 – at a distance of 310km from the base, on the beam of Sfintul-Gheorghe at the Danube (at 15km from the coastline) – the command to disconnect was given. One of the TB-3s launched the I-16s piloted by Captain Shubikov and Lieutenant Litvinchuk; the other one launched the I-16s piloted by Senior Lieutenant Filimonov and Lieutenant Kasparov. Within 15 minutes after the separation, the dive bombers approached the target and were met with heavy anti-aircraft fire. The four fighters broke through the fire, dived from 1,800m and released the bombs from an altitude of 300m; then, with no losses, flew away. At 0640, the I-16s were already

* Gatchina: collection 12; record statement 2, file 554, sheet 183.

kicking up dust at the Odessa aerodrome; however, on this occasion, Shubikov's four stayed on for some time in Odessa. The fighters took part in repelling the German bombers' attack on the city, and only then flew to Eupatoria.

The report of the raid's participants stated that the attack target was hit; however, it soon became clear that the bridge had survived and that the oil refinery had only been slightly damaged.

For this reason, the raid was repeated on 13 August – but this time by six fighters. The schedule was the same as for the previous sortie, with the only difference being the departure 30 minutes later. The attack took place at 0550; the pilots scored five direct hits on the bridge and completely destroyed one of the spans. On the way back, the fighters strafed the enemy's infantry near Sulina – and at 0705, they landed safely in Odessa; then they refueled and flew to Eupatoria.

On the same day (13 August), Design Engineer Vakhmistrov wrote in his letter addressed to the deputy head of the Directorate of Naval Aviation (Major General for Aviation Korobkov):

> As of today, SPB equipment has been reinstalled on three TB-3 aircraft and six I-16 aircraft of the 62nd Aviation Brigade; another Zveno aircraft will soon be complete. The following additional work has been performed throughout the period of my visit:
>
> 1. Torso back protective armor plates have been mounted to the seats of the left pilot, flight mechanic, and aft gunner.
> 2. In the aisle between the pilots' seats, a folding arch with a socket for installation of the dual DA gun has been installed; thus, the right pilot is enabled to keep the upper forward and upper rear hemisphere under fire.
> 3. An attachment fitting for suspension of a 95-liter droppable tank has been mounted under the I-16 fuselage. According to the new fueling scheme, after separation from the TB-3, the I-16s will be fueled from the droppable tanks; once the latter are dropped, the I-16s will be fueled from the main tanks. The droppable tank installation has been flight-tested, and has demonstrated failure-free operation.
>
> Throughout my stay here, Zveno-SPB has made two raids to Constanţa area; it should be noted that during the first raid the I-16s bombed the target at 13:00 in conditions of a clear sky. There was normal traffic in the city before the bombardment; the anti-aircraft artillery started firing only after the bombardment – this testifies to the fact that the appearance of our fighters over the target was quite sudden and unexpected. During the second raid [the bombing of the bridge over the Danube], 100% hits were scored, and the bridge was destroyed. In both cases, the aircraft returned intact. Such results of operational use of Zveno-SPB show that the configuration is highly efficient and should be used on a wider scale.
>
> Zveno-SPB will probably ensure considerable efficiency in raids against tank columns. In this case, i.e. at smaller distances, a single TB-3 may service three or four pairs of the I-16s and lift them to an altitude of 400 to 500m pair after pair.
>
> The armoring and fire weapons improvement work that I have completed on the TB-3 is not sufficient. There is a possibility to add two more gun posts in the bomb hatches for firing in the forward and aft-downward direction. Besides, it would be very desirable to provide large-caliber weapons in the navigator's turret and in the aft gunner's turret.
>
> The armor which I have installed was taken from the I-15 aircraft; however, special armor plates are required.
>
> The fuel tanks, which have no protection against fire, are the most vulnerable components on the TB-3. It is desirable to promptly raise the issue of such protection; this will probably not be connected with any special difficulties.

> To summarize my points:
> 1. Increasing the number of SPBs (10 TB-3s and 40 I-16s at the Black Sea),
> 2. Increasing the number of gun posts and the mounting of large-caliber guns on the TB-3,
> 3. Installing special armor plates to the TB-3,
> 4. Protecting the TB-3's fuel tanks and supplying inert gas to them.
>
> The Black Sea Air Force Command ordered that I should report the above to you for further report to People's Commissar of the Navy Comrade Kuznetsov and for implementation.*

The letter of Major General Korobkov to the People's Commissar of the USSR Navy – Admiral Kuznetsov – dated 13 August 1941 raises the issue of an additional order for the manufacture of equipment for the assembly of subsequent 'Zveno Aircraft'. In view of expected problems with the manufacture, since the most suitable Aviation Factory No.207 is still busy manufacturing the Su-2s, General Korobkov suggests focusing on the opportunity to use other dive aircraft:

> The Naval Air Force has, at its disposal, twelve TB-3 4A34RNs; of these, five aircraft are already fitted with engineer Vakhmistrov's suspensions. I believe that it would be more practical to use the remaining seven aircraft as transport ones. And we should not expect receiving even ten TB-3 4A34RNs from the Red Army Air Force, because manufacture of such aircraft was discontinued in 1937, and they had been manufactured in the limited quantity of approximately 150 aircraft. Currently, the combatant forces are beginning to receive the Petlyakov Pe-2 dive bombers; these, once they are mastered and used in accordance with their intended purpose, will provide better results than the I-16 SPB fighters as used in engineer Vakhmistrov's design.†

Indeed, after the outbreak of war, the Black Sea Fleet received Pe-2 dive bombers which were mastered by pilots of the 40th Bomber Regiment. On 1 August 1941, six Pe-2s from the 40th Bomber Regiment made their first attack on the ships in Constanţa port. According to the combat report, a transport vessel and a tugboat were destroyed. On their way back to the base, the Pe-2s were attacked by German fighters, but managed to repel the attack and return to the base with no losses. Later sorties demonstrated the rather high efficiency of the Pe-2s as well: on 10 August, they took part in the Cernavodă bridge raid, together with the I-16 SPBs; it is understandable, therefore, that many fleet commanders believed the Pe-2s more efficient in terms of dive-bombing.

Meanwhile, People's Commander Kuznetsov understood that the fleet would probably receive the extra Pe-2s from the industry, but there was no reason to abandon the idea of assembling additional 'Zveno Aircraft' as well.

With this in mind, on 16 August 1941, Kuznetsov sent the following letter to Joseph Stalin:

> To: Chairman of the Council of People's Commissars of the USSR Comrade J.V. STALIN
>
> During the period from 26 July thru 13 August, the combined Zveno aircraft (a TB-3 + two I-16s) designed by engineer V.S. Vakhmistrov were combat-tested in the bombing raid on the Constanţa port oil depot and the Cernavodă bridge over the Danube. According to the report by the Black Fleet Commander Vice Admiral Oktyabrsky, the destruction of the Cernavodă bridge is well attributable to the use of engineer Vakhmistrov's system, which allows attaching, to a TB-3 aircraft, two

* Gatchina: collection 12; record statement 2, file 543, sheets 86-87.

† Gatchina: collection 12; record statement 2, file 543, sheets 89-90.

I-16s carrying two FAB-250 bombs each. The combined Zveno aircraft consisting of a TB-3 and two I-16s may be used for the following purposes:

1. Bombing of small-size stationary targets and the enemy's ships; and
2. Loitering to ensure anti-aircraft defense of fleet bases and laying vessels.

At the same time, it should be noted that the combined Zveno aircraft (a TB-3 + two I-16s) have the following drawbacks:

1. Taking into account modern fighters' armament, the TB-3 aircraft is poorly armed (7.62 caliber machine guns) and has no armor protection; fuel tanks are not protected either.
2. The TB-3 aircraft were manufactured in 1936-37, and are considerably worn; this fact brings about certain limitations in terms of their gross weight. Currently, the Naval Air Force has, at its disposal at the Black Sea, only five TB-3 aircraft and nine I-16 aircraft with special equipment. This is not enough to carry out raids on the required scale and to obtain any tangible results.

Please authorize preparation of additional ten combined Zveno TB-3 aircraft with 20 I-16s; for this purpose, it is required that the Red Army Air Force provide ten TB-3 4AM-34RN aircraft and twenty I-16 Type 24 aircraft. It is necessary that steps as set forth by me in the Draft Resolution of the Council of People's Commissars be implemented; for this purpose, attached please find the said Draft for your approval.

Attachment: Draft Resolution of the Council of People's Commissars on two sheets

Signed: People's Commissar of the USSR Navy Admiral Kuznetsov*

Indeed, the Draft Resolution of the Council of People's Commissars was prepared and requested that the 'Head of the Red Army Air Force provide ten TB-3 4AM-34RN bombers and 20 I-16 Type 24 fighters before 01 September 1941'. The People's Commissariat for Aviation Industry was to manufacture the required equipment and to fully equip ten Zveno-SPB systems before 15 October 1941. At the same time, it was suggested that before 15 September, the 'possibility to use the TB-7 aircraft as a carrier for modern fighters in accordance with engineer Vakhmistrov's design should be checked and reported'.†

In connection with the latter suggestion, a letter was sent on 9 September 1941 on behalf of the deputy head of the Naval Air Force – Major General Korobkov – to Aviation Factory No.124 in Kazan, where the TB-7 bombers were built. Addressed to Factory Director Akulov and Chief Designer Nezval, the letter raised the issue of the possible use of the TB-7 as a carrier for a pair of Mikoyan-Gurevich (MiG-3) or Lavochkin-Gorbunov-Gudkov (LaGG-3) fighters. The letter read: 'Is it possible, in terms of design and taking into account the dimensions and strength performance of the TB-7, to attach, under its wing, two MiG-3 or LaGG-3 fighters carrying, along with conventional armaments, two FAB-250 bombs? What principal modifications of the TB-7 would the attachment of the fighters entail?' Furthermore, the letter requested the estimated flight performance of such Zveno aircraft and the agreement to modify the TB-7 for attachment of the fighters during the current year.‡

The letter on behalf of Akulov and Nezval followed in respect of the use of the MiG-3 only, because Factory No.124 did not have the required data on the LaGG fighter at its disposal:

1. In terms of dimensions and strength, the TB-7 permits the attachment of two MiG-3 type fighters carrying a bomb load of two FAB-250s each. In such a case, design modifications to

---

* Gatchina: collection 12; record statement 2, file 555, sheets 282-283.

† Gatchina: collection 12; record statement 2, file 555, sheet 284.

‡ Central Archive of the Navy, Gatchina: collection 12; record statement 2, file 543, sheets 96-97.

the TB-7's wing would include, along with installation of two special beams, reinforcement of two wing ribs at the fighter attachment points by means of replacement of the existing braces and their attachment fittings between the spars with stronger ones.

2. Estimated flight performance of a Zveno configuration comprising a Tupolev TB-7 bomber + two Mikoyan-Gurevich MiG-3 fighters:

Weight performance in terms of maximum allowable take-off load:

	AM-35A	M-40
TB-7 operational empty weight (with the crew, weapons, and ordnance) (kg)	20,355	21,955
Normal gross weight of two MiG-3 aircraft (kg)	6,200	6,200
Bomb load of the MiG-3 aircraft (kg)	1,000	1,000
Fuel and oil weight (kg)	5,945	4,345
Maximum allowable gross weight (kg)	33,500	33,500

1. Level flight speeds at H=6,000-7,000m with all six engines operating at rated setting and with an average gross weight of 30,000kg: 420km/h with AM-35A engines, and 405km/h with M-40 engines.
2. Rate of climb and service ceiling:

	AM-35A	M-40
Time to climb to 6,000m with weight of 33,500kg (minutes)	18.0	18.5
Service ceiling with weight of 33,500kg (m)	8,700	8,900

3. Zveno aircraft take-off performance:

	AM-35A	M-40
Take-off run with all six engines boosted, take-off weight of 33,500kg, and no landing flaps used (m)	520	500

4. Range and endurance at an altitude of 5,000-6,000m and a speed corresponding to minimum fuel consumption, but taking into account the fighters' refueling from the TB-7's tanks:
5.

	AM-35A	M-40
Zveno aircraft range (km)	1,450	1,320
Duration (h)	5.0	4.5
Remaining fuel in TB-7's tanks (kg)	500	500

Note: Currently, the possibility to increase the initial gross weight to 34 tons is being checked experimentally. If this turns out to be possible, the range will be increased to 1,620km with AM-35A engines and to 1,530km with M-40 engines.

6. In view of the fact that the factory is busy fulfilling other orders, the modification in question cannot be implemented during the current year.*

Thus, the development of Zveno-SPB configuration at this stage was terminated entirely; however, operational use of the existing TB-3 + two I-16 systems continued. On 17 August, six I-16s successfully destroyed a dry dock in Constanţa. At the end of the month, the dive bombers were used to destroy crossings at the Dnieper River. The importance of the mission comprised the strategic value of the crossings and the possibility to considerably disrupt the enemy's advance. On 28 August, two TB-3s took off from the Eupatoria aerodrome and headed for the Zaporozhye area. At dawn – at a distance of 30 kilometers from the city – the I-16s disconnected from the carrier. The attack was sudden and accurate; all four fighters left with no losses. On 8 September, a Zveno SPB – in cooperation with the Yakovlev Yak-1 escorting fighters – destroyed a crossing near Berislav; a Yak-1 and one of the I-16s were lost. On the next day, after a successful bombing raid, four I-16s shot down two Messerschmitt Bf. 109s.

Further operational use of the 'suspended aircraft' took place under increasingly aggravated conditions for the Soviets' military situation. In late September, German troops started to advance in the northern part of the Crimea – and all available aircraft (even those intended for high-precision strikes in the enemy's rear) were used to hold the enemy back. Under the circumstances, the TB-3s were often used only to lift the I-16s armed with heavy bomb loads, because the fighters could not get airborne on their own carrying two 250kg bombs each. Generally, bombing the enemy's artillery batteries with 250kg bombs usually resulted in destroying said batteries.

The hard combat conditions resulted in an increase in casualties. On 1 October 1941, the commander of the squadron of suspended aircraft – Arseny Shubikov – was killed in an air engagement; later on, the suspended I-16s were only used episodically, and combat reports became scarcer. Only one thing is known for certain: in total, Zveno SPB flew at least 30 combat missions, which, due to their high efficiency, constituted one of the most successful military efforts of the Soviet aviation during World War II.

* Central Archive of the Navy, Gatchina: collection 12; record statement 2, file 552, sheets 286-287.

The List of Principal Work on Combined Aircraft in 1931-1941: The Table Prepared by V.S. Vakhmistrov himself*

No.	Designation	Date Devised	Design Description	Principal Available Documents	Note
1.	Zveno-1 (Z-1) TB-1 + two I-4s	03 December 1931 thru July 1933	Two I-4s were mounted to the TB-1 with their lower wings removed, oil tanks added, and attachment locks installed. The TB-1 was fitted with 'spiders' accommodating forward locks, which were controlled from the copilot's seat. The landing gear was reinforced, and the fuel supply system modified	1. Revolutionary Military Council's Report on Z-1 first flight, dated 05 December 1931 2. Report on completion of flight tests of 01-08 April 1932 3. Report on operational tests of August 1933 4. P.M. Stefanovsky's individual flight record	After the first flight, the release locks were modified. A small batch of Z-1s was manufactured for operational tests; the tests were successful
2.	Z-1a TB-1 + two I-5s	September 1933 thru 05 October 1933	The TB-1 was modified for the mounting of the I-5s	1. Z-1a test report of September 1933 2. Z-1a test report as approved by Alksnis, dated 05 October 1933	A small batch of Z-1as was manufactured. The operational tests were successful
3.	Tyagach Zveno (T-Z)	1933	A special-purpose aircraft for acceleration of the carrier aircraft take-off, with a powerful engine and a specially selected propeller	Preliminary design	The work never went further than the preliminary design
4	Z-2 TB-3 + two I-5s	June 1933 thru August 1935	The design of attachments for the three I-5s was similar to that used on Z-1a	1. P.M. Stefanovsky's individual flight record 2. Report on tactical tests of Zveno 'TB-3 + two I-5s' aircraft (with refueling)	Configurations of 'a TB-3 + two I-5s', 'a TB-3 + an I-5', and 'a TB-3 + three I-5s' were tested
5.	Z-3 TB-3 + two I-Zets	July 1934	Two I-Zets were attaching using a special floating frame. During the TB-3's take-off, the I-Zets were locked in the extreme lower position	P.M. Stefanovsky's individual flight record	The first experimental flight ended with a crash, killing pilot Korotkov. The work was terminated entirely

* A collection of documents relating to V.S. Vakhmistrov, file 5196/149: a summary table of V.S. Vakhmistrov's combined aircraft designs.

No.	Designation	Date Devised	Design Description	Principal Available Documents	Note
6.	Z-4 TB-3 + I-5 (fuselage)	November 1935	A simplified T-Z design utilizing an I-5 fuselage without wing and empennage		Several flights; no advantages found; flights terminated
7.	Z-5 TB-3 + I-Zet	04 April 1935	The TB-3 was provided with a truss for attachment of the I-Zet	1. Test report 2. P.M. Stefanovsky's individual flight record	With participation of pilot V.A. Stepanchonok
8.	Z-6 TB-3 + two I-16s	December 1934 thru 13 December 1936	The I-16s were attached to the TB-3 in three points each. The pyramid vertices accommodated locks. The third attachment point was located behind the I-16's pilot seat; the brace was reclining	1. P.M. Stefanovsky's individual flight record 2. Flight test report	The I-16 taxied up to the TB-3, attached to the locks, and retracted the landing gear. The aft lock was modified to exclude self-release. The tests provided positive results. System for the refueling of the I-16s from the TB-3 was modified
9.	Zveno Fighter	1934-1935	High speed at the design altitude; small wing area; no conventional landing gear; return to carrier after the mission	A memo report to Deputy Chairman of the Revolutionary Military Council of the USSR Mikhail Tukhachevsky and the latter's resolution to include the Zveno Fighter in the Work Plan	The aircraft was designed and entered production at Factory No.22, but then discontinued and removed from the Plan
10.	Aviamatka PVO (A-M) TB-3 + two I-5s + two I-16s + I-Zet	1935	Z-2, Z-5 and Z-6 combined and tested in accordance with a special program in cooperation with Factory No.22	1. Report on experimental tests of 'TB-3 + two I-5s + two I-16s + an I-Zet' configuration 2. P.M. Stefanovsky's individual flight record	Two training flights and one test flight with positive results
11.	'SPB I-16' TB-3 + two I-16s carrying two FAB-250 each	1936-1940	FAB-250 bombs were suspended on special locks under the beam, to which the I-16s were attached	1. A suggestion of the Revolutionary Military Council dated 14 August 1936 (on the occasion of the Day of the Navy) 2. 1938 flight test report 3. A suggestion to use in the Soviet-Finnish War	Low-scale mass production; a single squadron in Eupatoria; participation in World War II

No.	Designation	Date Devised	Design Description	Principal Available Documents	Note
12.	SPB FAB-500	1941	A special frame was used for attachment of FAB-500 bombs. It allowed the bombs to avoid contact with the rotating propeller disc		One aircraft was equipped, but tests were never conducted
13.	Z-7 Two I-16s docking in flight	1939	Design of trusses – same as on Z-5; retaining pyramids	I-16 flight test report	Several flights for the training of pilots; P.M. Stefanovsky and S.P. Suprun; Tests not completed because of the outbreak of war

13

Later Projects by Vladimir Vakhmistrov before 1945

TRANSPORT GLIDER AND 'Σ' (SIGMA) MOTORIZED GLIDER

At this point of our story, let us remember that in mid-1940, all research work on Vakhmistrov's combined aircraft was officially terminated. At almost the same time, mass-manufactured Zveno-SPBs were completed and delivered to the customer (the Black Sea Fleet) in August and September 1940. All this meant that the current part of the story had come to an end, and designer Vakhmistrov now had to look for new areas of activity (without, though, abandoning his primary topic of combined aircraft).

During the period in question, Vakhmistrov addressed a letter to the Deputy People's Commissar of Aviation Industry in Experimental Aircraft Building A.S. Yakovlev, Air Force Commander Y.V. Smushkevich, the head of the Naval Aviation Directorate S.F. Zhavoronkov and Air Force Commander A.I. Filin. Vakhmistrov began his letter with general ideas regarding the methods of waging the war, which had already been going on, with a special emphasis on actions of troop-carrying aircraft and dive bombers. The troop-carrying aircraft would soon be reflected in the designer's work in the form of the combat troop-carrying glider (*BDP*) project, which we have already discussed above. As for dive bombers, Vakhmistrov adhered to his traditional concept and suggested using, for the purpose, conventional fighters to be suspended under a large, specially designed troop-carrying aircraft. Furthermore, he wrote that he was ready to take part in the designing of such troop-carrying aircraft.

The above extracts of the letter were found in the unsorted folders of Vakhmistrov's archive; they bear neither dates, nor output data, and may well be pieces of a draft letter; yet, four years later, the inventor returned to his combined aircraft in a somewhat modified form. This time, he presented for review a flying vehicle which was described as a glider, or a 'Σ' (Sigma) Motorized Glider. By all appearances, it was one of the versions of the troop-carrying glider as devised before the war – and apparently, the modification was inspired by the successful use of transport and troop-carrying gliders during the war. Among the many versions of improvements for such gliders, special attention was given to the idea of providing them with engines – thus converting them into Motorized Gliders. In particular, in 1943, Polikarpov's combat troop-carrying glider was modified in such a manner. It was fitted with two M-11F engines, which helped it to continue straight flight without descent after separation from the towing airplane. In practice though, the Motorized Glider was never used; however, the Transport Gliders were, as such, used quite successfully in experimental cargo transportations in the USSR's rear. In particular, the possibilities of using the gliders in the national economy were studied (along with the Civil Air Fleet Scientific Research Institute) by the Experimental Glider Branch of the Moscow Civil Air Fleet Directorate, which had, at its disposal, G-11 gliders. Specialists of the Civil Air Fleet Scientific Research Institute claimed that such 'glider trains' were highly efficient.

Vakhmistrov, who had, at the time, a keen interest in configurations for air cargo transportation using gliders, suggested another original project: initially, it was a large Transport Glider Σ.T.G. (TG+2 fighters) with a wing span of over 33 meters. Two fighters, type Yak-1 or Yak-9 (each of which had a gross weight of 3,000kg, a wing area of 17.5m^2 and an engine power of 1,630hp), were attached under the wing of this high-wing glider – thus, the suggested system was, to a large extent, similar to the already known 'Zveno Aircraft'. The fighters had a high power-to-weight ratio and could lift the glider into the air and tow it to a target at a distance of more than 1,000km.

The next version in this project was the Motorized Glider Σ.M.G. (MG+2 fighters), with a wing span of more than 40m, two ASh-82 engines and (like in the previous version) two under-wing fighters. Both projects – the Transport Glider and the Motorized Glider – were approved by Vakhmistrov on 9 July 1945. It should be noted that the idea as it is, given the clarity of the concept, was a colossal one, because the Transport Glider and the Motorized Glider were rather large flying vehicles. Suffice it to say that the Soviets' largest troop-carrying glider – IL-32 (built in 1948) – had a wing span slightly more than 35m; however, the above-described 'Σ' (Sigma) project was never implemented in practice.*

Principal Performance Parameters

	Transport Glider Σ.T.G. (TG+2 fighters)	**Motorized Glider Σ.M.G. (MG+2 fighters)**
Wing span (m)	33.60	40.00
Length in the flight line (m)	20.00	24.00
Gross weight (kg)	10,660	26,000
Gross weight with two fighters (kg)	16,660	32,000
Payload (kg)	5,330	12,800
Maximum speed with two fighters (km/h)	306	340
Service ceiling (m)	7,000	6,000
Range (km)	1,200	960

COMBINED AIRCRAFT PROJECT

As late as at the final stage of World War II, the Red Army Air Force again turned its attention to such flying vehicles as flying bombs, or cruise missiles. One of the reasons for the revived interest was the use of such flying vehicles by the air force of fascist Germany; Vakhmistrov did not disregard the topic either. In 1944, he filed in an application for an invention referred to as the 'Aircraft for Glider Towing (A Combined Aircraft)'. The glider was of quite an original twin-boom design, which successfully allowed it to connect to the aircraft (a fighter). In this configuration, the aircraft was mounted on top of the glider in a 'rider' fashion, and its engine thrust was used to make the combination airborne. The landing gear – connected to the glider – was intended to be dropped after take-off. Once close to the target, the glider disconnected and could further be used as a flying bomb. The concept of the combined aircraft was, though rather extravagant, quite implementable. It was because of those several projects that the management of the *NII VVS*, which had been cooperating with Vakhmistrov for many years, suggested taking an academic degree without the need to defend a thesis paper.

On 31 March 1944, the head of the *NII VVS* – Lieutenant General Losyukov – sent a letter to the Higher Education Committee, attached to the Council of People's Commissars of the USSR, which read:

> During the period from 1931 thru 1941, engineer V.S. Vakhmistrov made a lot of work to develop his concept of 'combined aircraft'.

---

* A Branch of Russian State Archive of Scientific and Technical Documentation: Collection r-1; record statement 38-5, file 1418, sheet 5. State Research Institute of Civil Aviation (*GosNIIGA*), 1944. V.S. Vakhmistrov: 'A Layout of Transport Glider and Motorized Glider with Under-Wing Fighters (Zveno Type Configurations)'.

As an engineer for the Red Army's NII VVS, Comrade Vakhmistrov suggested, in 1931, mounting two Polikarpov I-4 fighters on top of the Tupolev TB-3 bomber's wing. Comrade Vakhmistrov substantiated his idea by theoretical calculations and wind tunnel tests, and brought his suggestion to the implementation stage.

The flight tests confirmed all the calculations in full.

In his further work, Comrade Vakhmistrov kept developing his idea; finally, in autumn 1941, the combination of a TB-3 and two I-16s armed with FAB-250 bombs was very efficiently used in military activities against the Romanian aggressors.

In view of the above, the Red Army's NII VVS hereby requests the Higher Education Committee attached to the Council of People's Commissars of the USSR to confer the academic degree of the Candidate of Science (Engineering) on Vladimir Sergeyevich Vakhmistrov without the need for the latter to defend a thesis paper.*

* A Branch of Russian State Archive of Scientific and Technical Documentation, Samara: Collection r-1; record statement 51-5, file 709, sheet 3. 'An Aircraft for Glider Towing (A Combined Aircraft): An application for an invention, 1944'.

Appendix: Scale Drawings

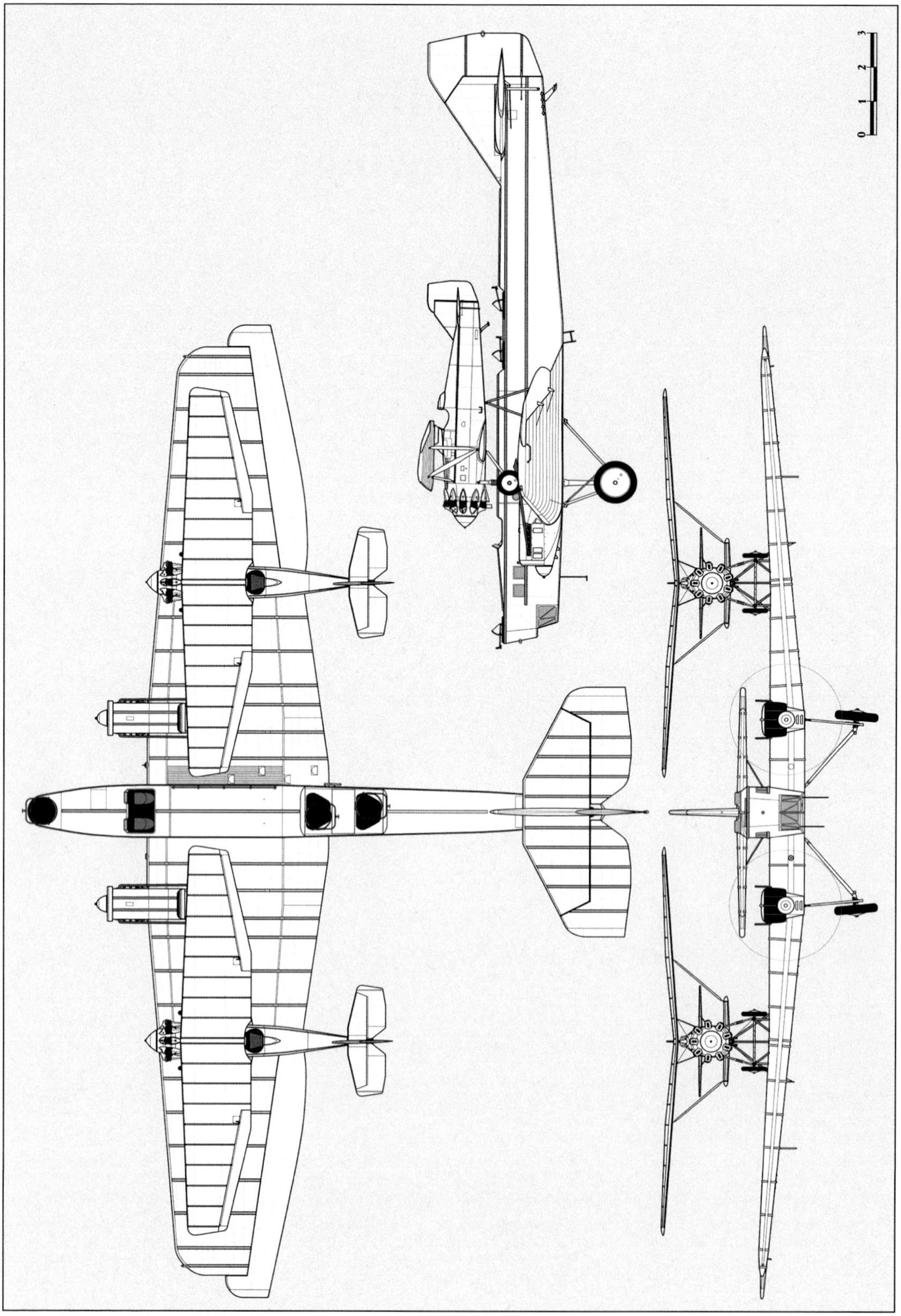

Zveno-1.

Zveno-1a.

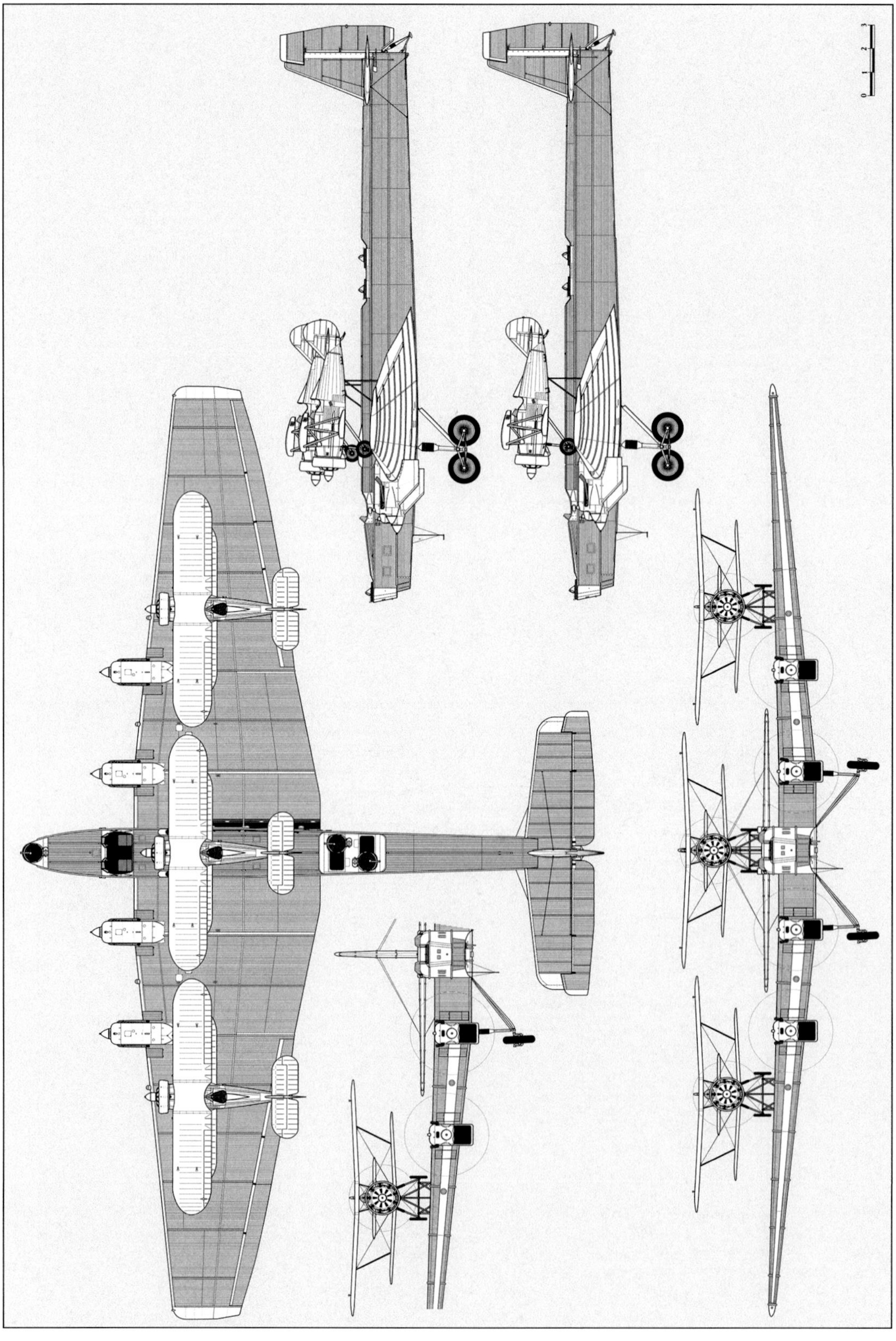

Zveno-2 (three views) – full configuration: TB-3 4M-17 + three I-5s, 1933; Zveno-2 (bottom side view and partial front view) – TB-3 4M-17 + two I-5s, 1935.

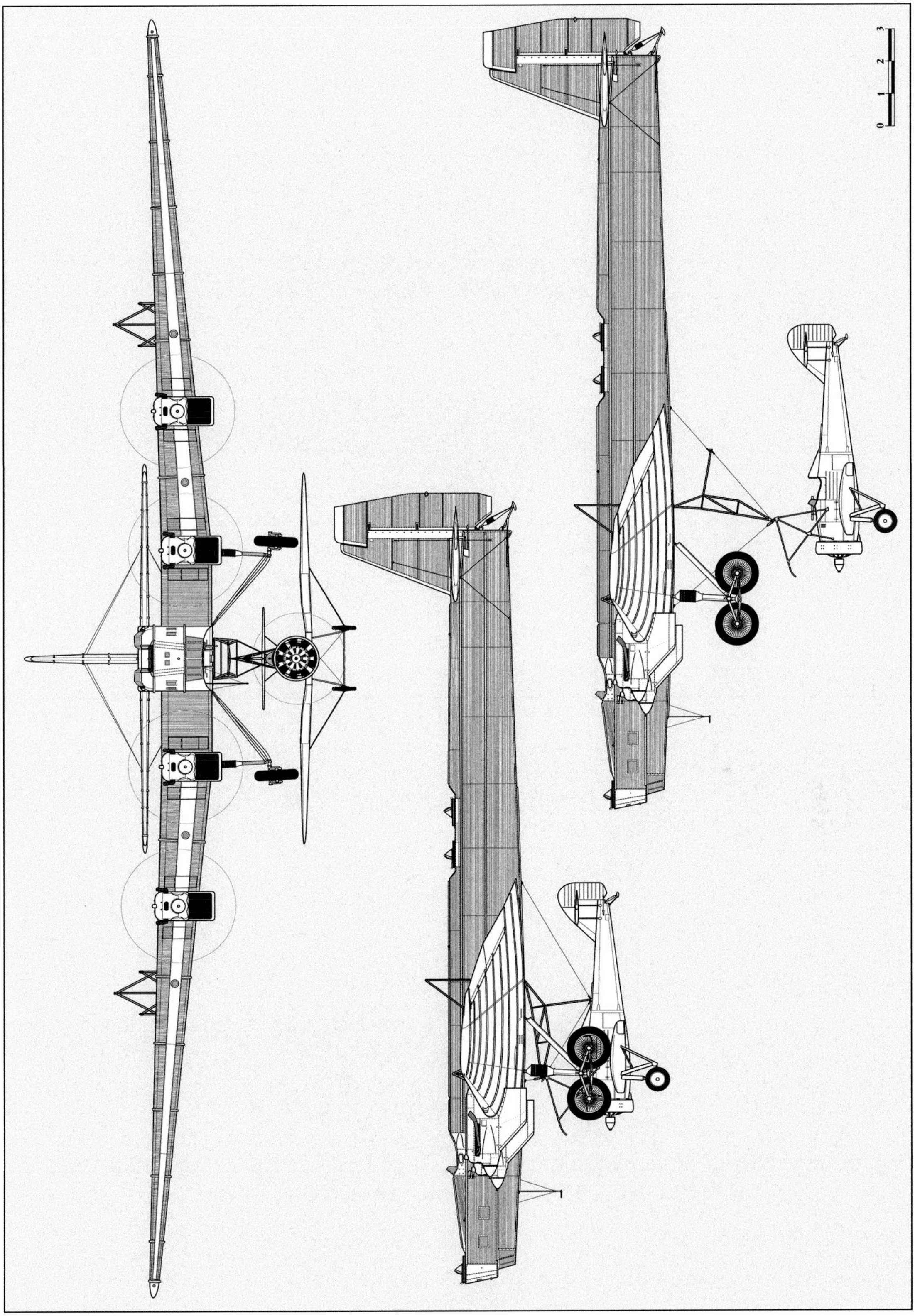

Zveno-5 – TB-3 4M-17 + I-Zet.

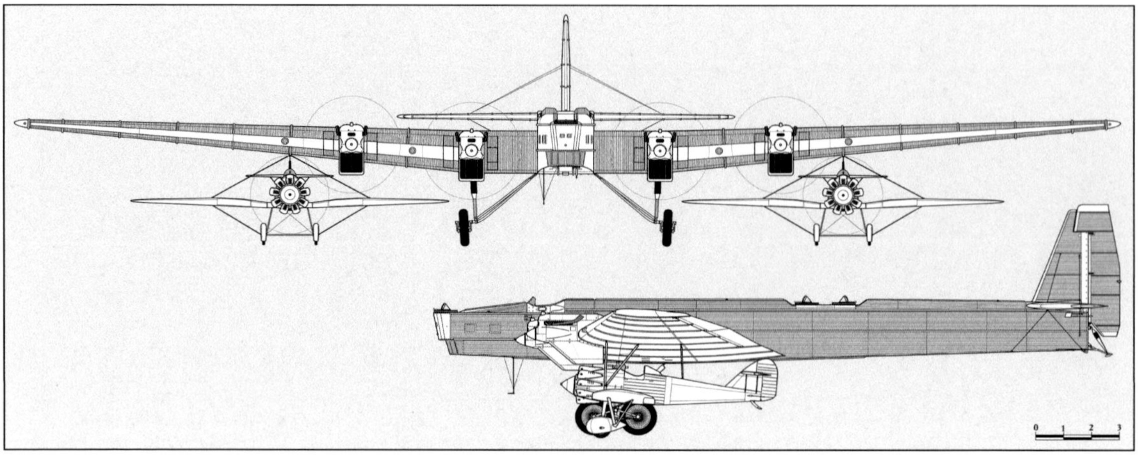

Zveno-3 – TB-3 4M-17 + two I-Zets.

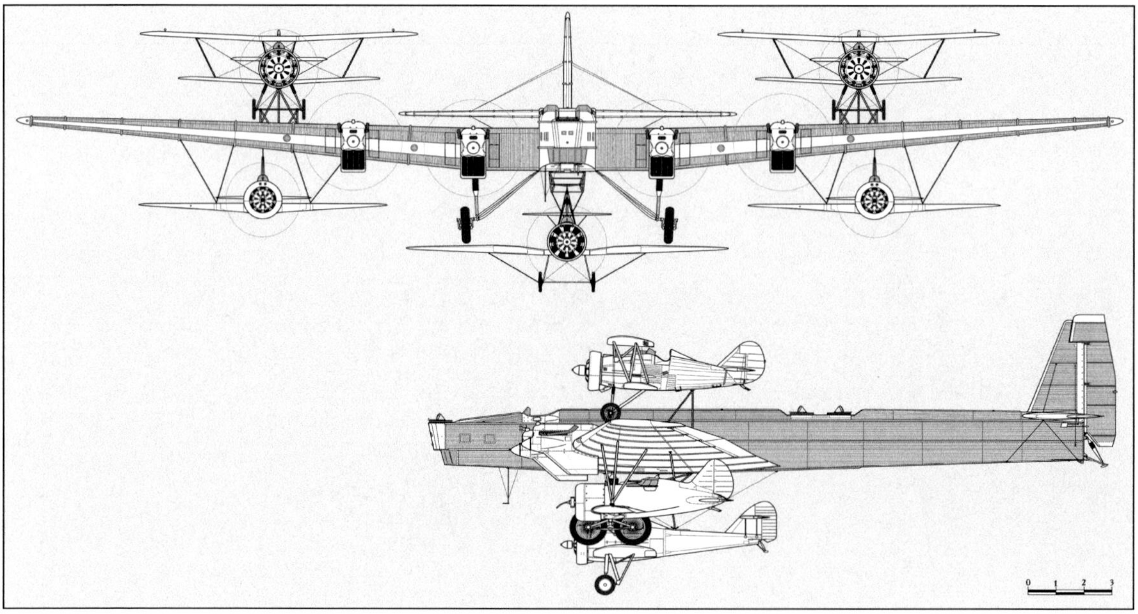

Aviamatka PVO – TB-3 4M-17 + two I-5s + two I-16s + I-Zet.

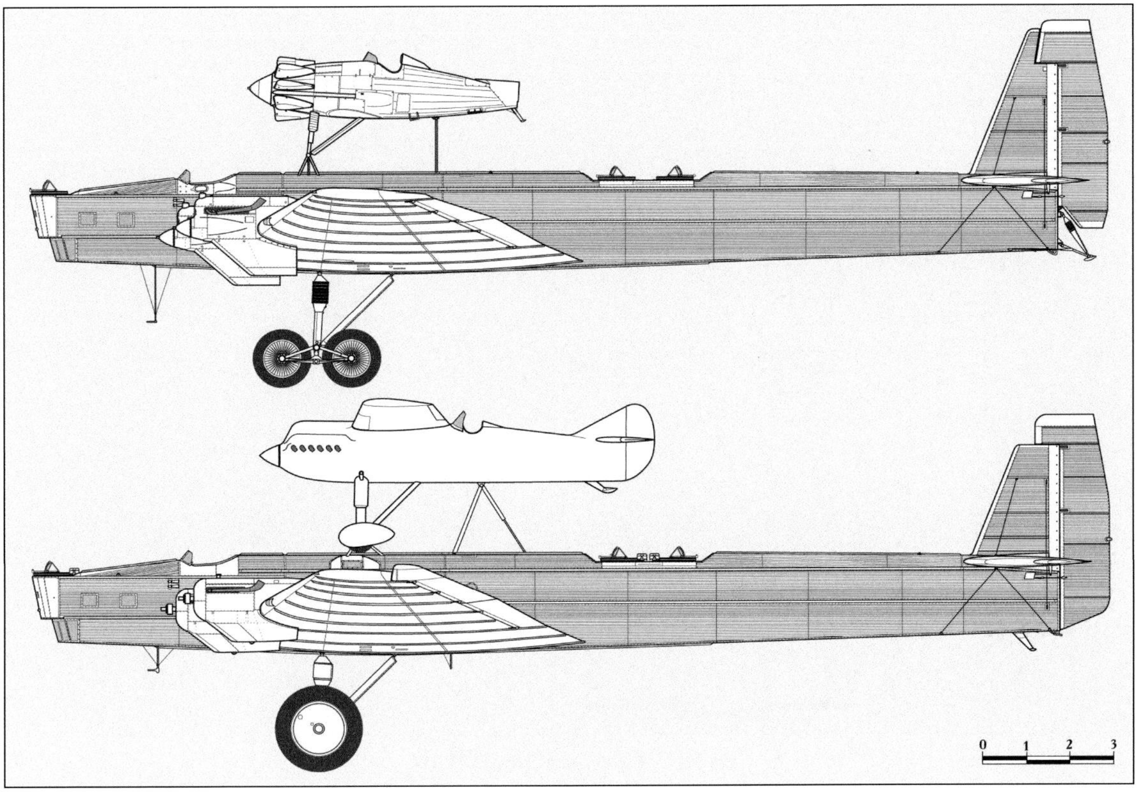

Tyagach Zveno (upper view) – TB-3 4M-17 + I-5 fuselage; Zveno-4 (bottom view) – TB-3 with T-Z aircraft.

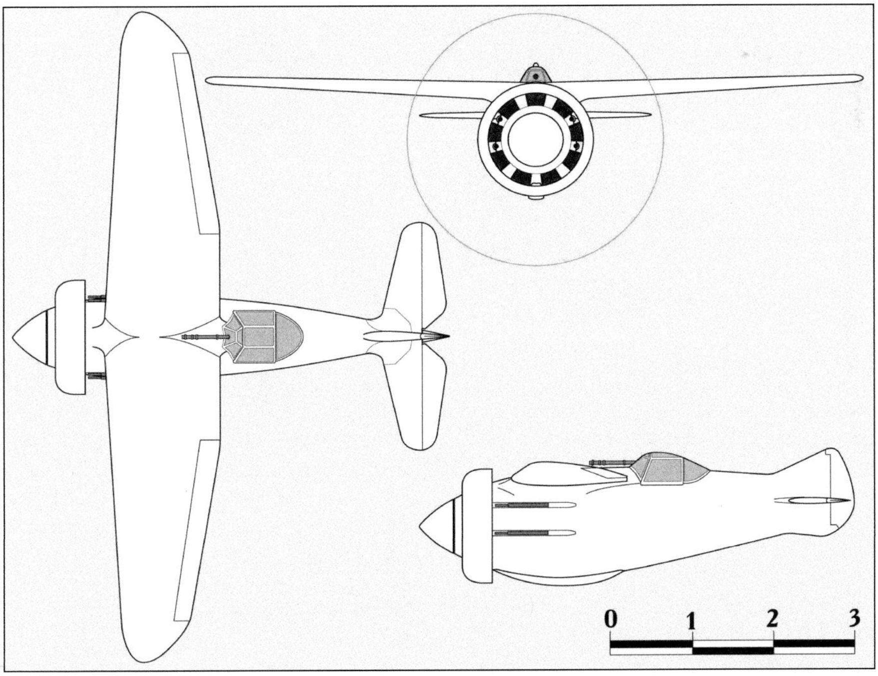

Project of Zveno Fighter (1933).

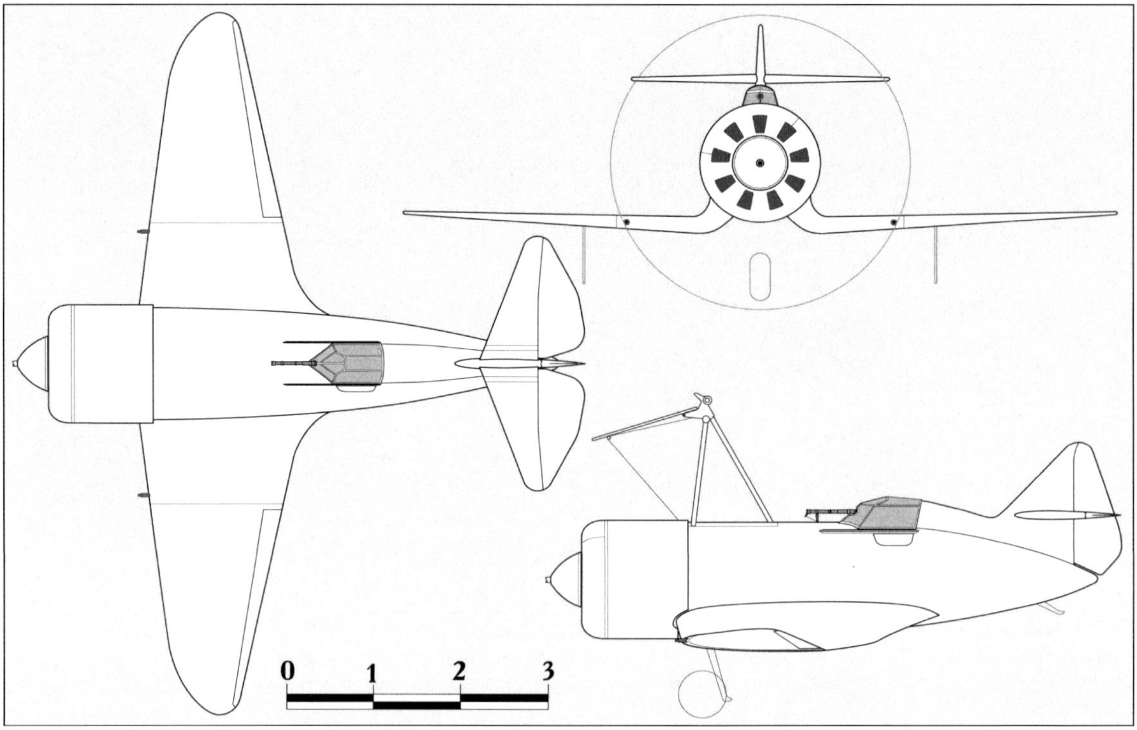

Project of Zveno Fighter (1935).

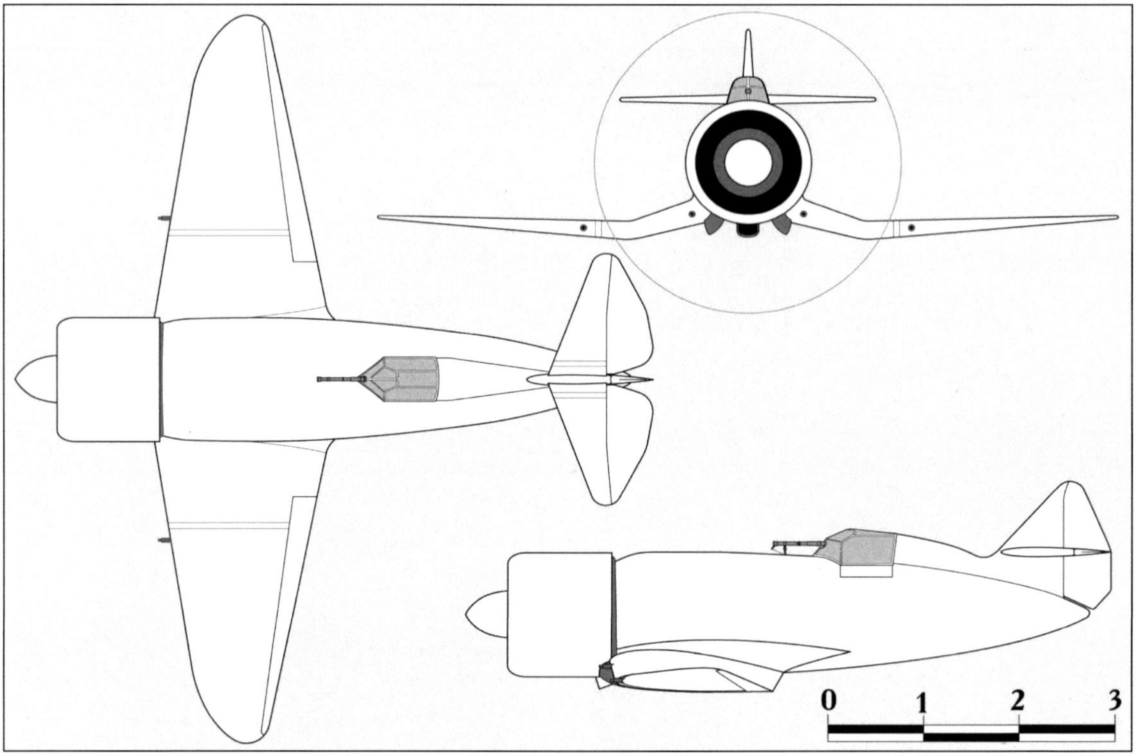

Project of the 'Airplane D'.

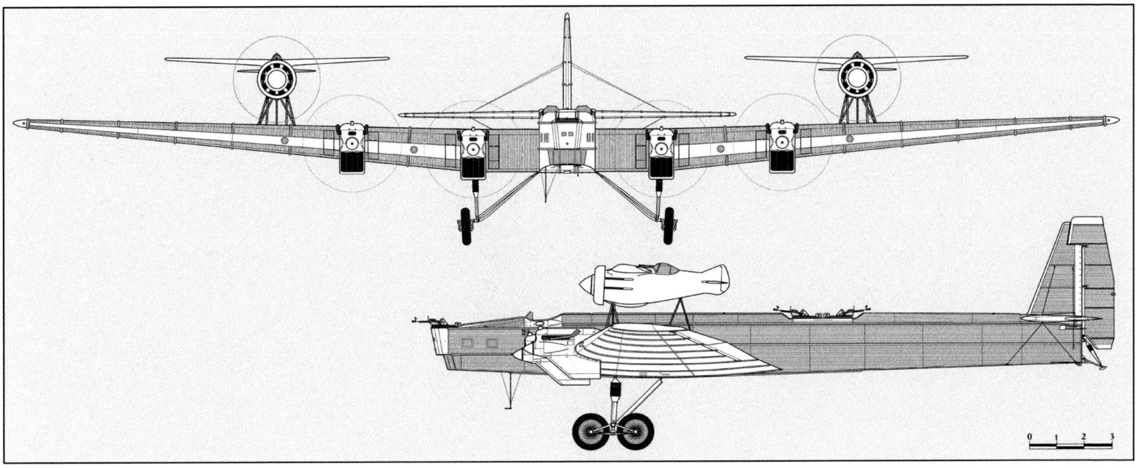

Zveno project with Zveno Fighters (1933).

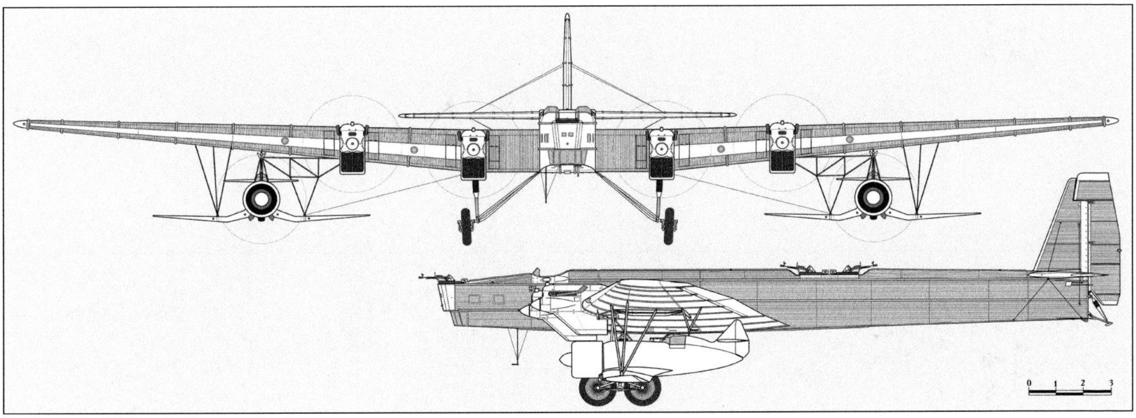

Zveno project with 'Airplane D'.

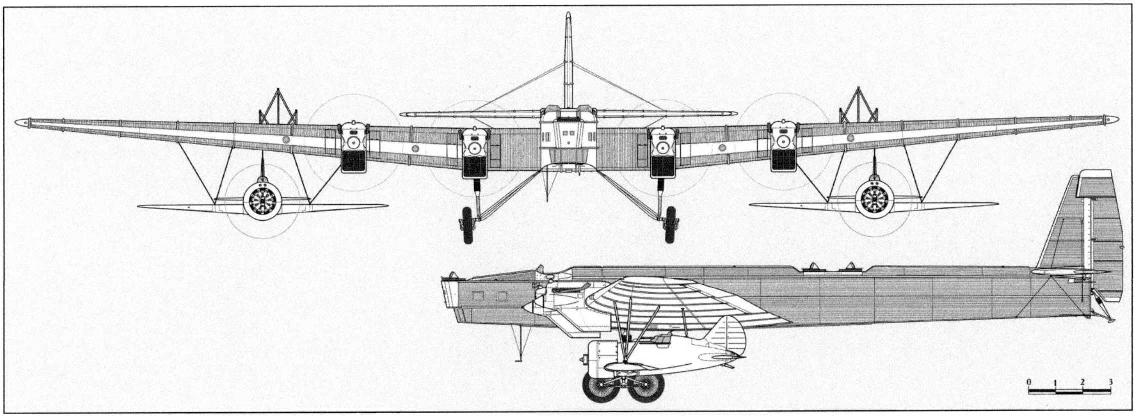

Zveno-6 – TB-3 4M-17 + two I-16s Type 4.

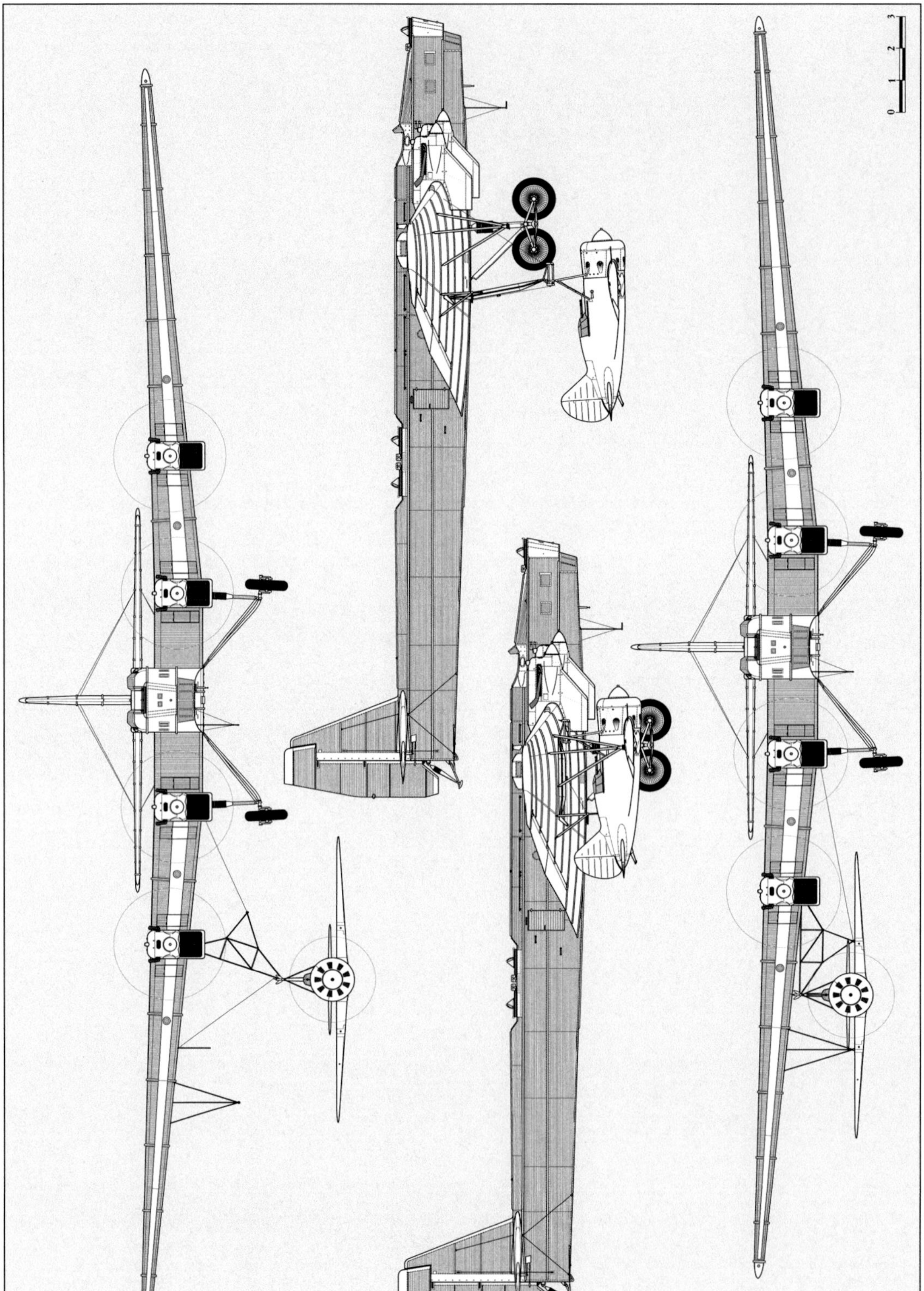

Zveno-7-1 – TB-3 4M-17 + I-16 Type 5.

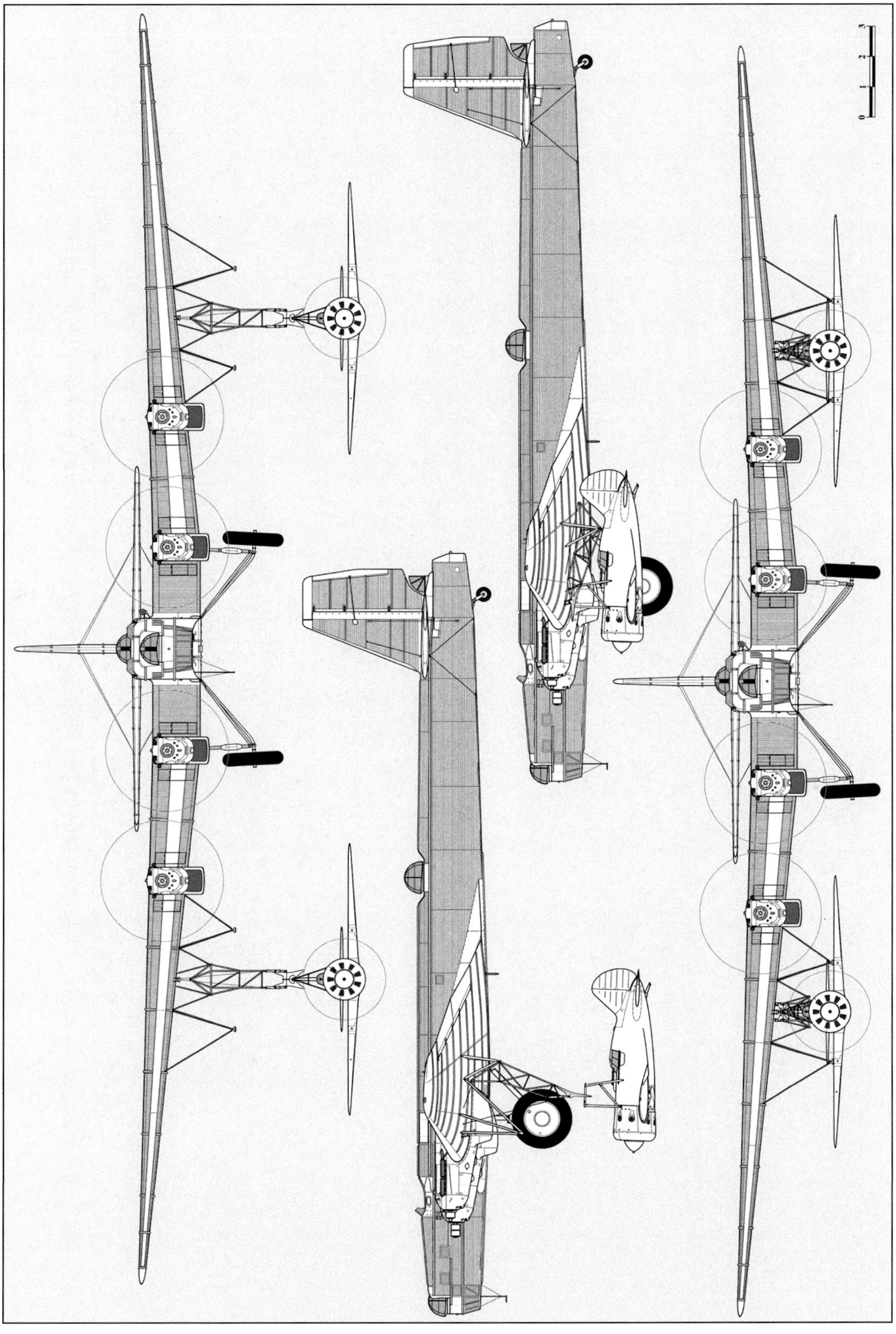

Zveno-7 – TB-3 4M-17 + two I-16s Type 5.

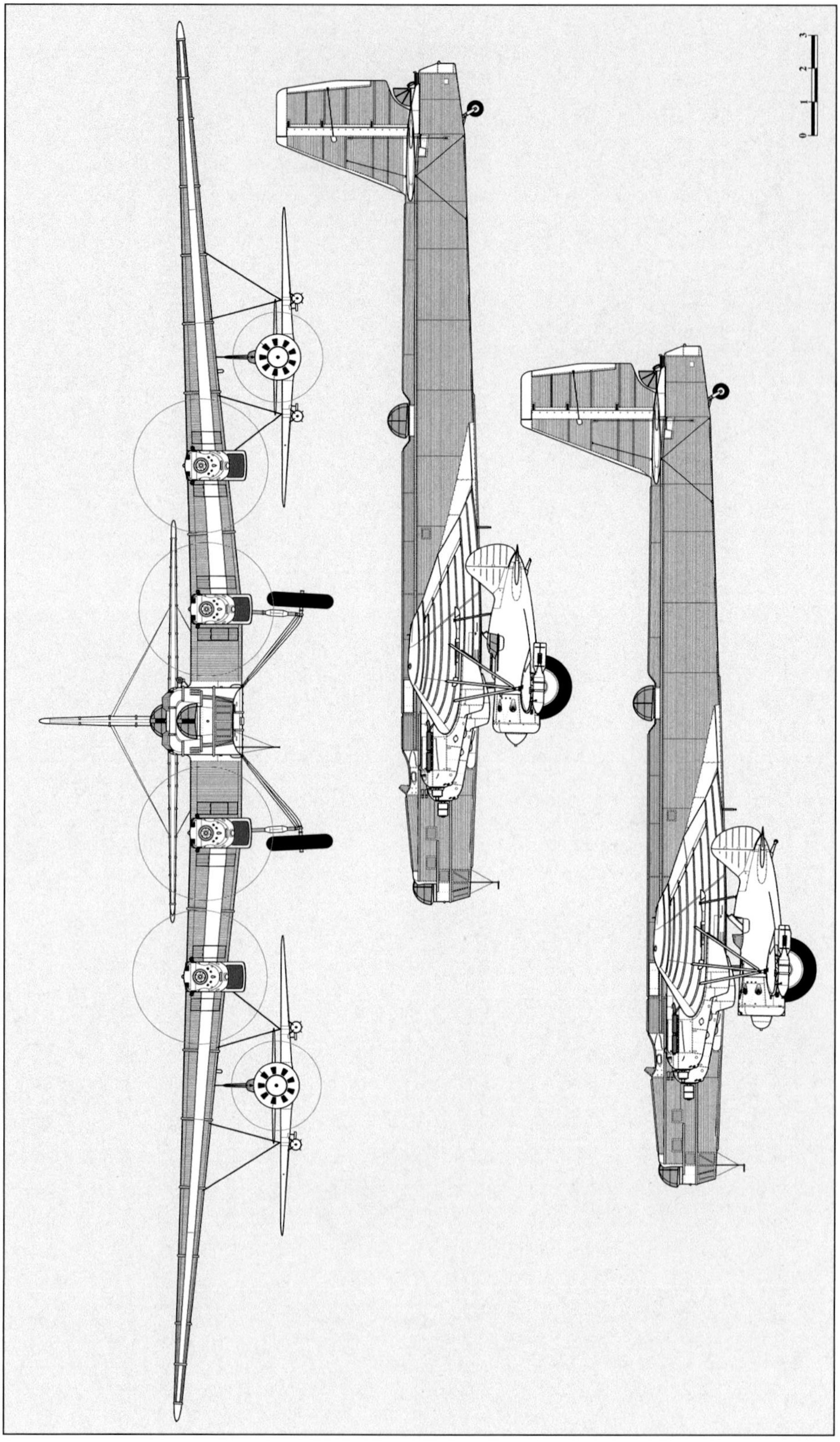

Zveno-SPB (upper front and side views) – TB-3 4M-34RN + two I-16s Type 5; Zveno-SPB (lower side view) – TB-3 4M-34RN + two I-16s Type 24.